Combinatorial
Knot Theory

SERIES ON KNOTS AND EVERYTHING

ISSN: 0219-9769

Editor-in-charge: Louis H. Kauffman *(Univ. of Illinois, Chicago)*

The Series on Knots and Everything: is a book series polarized around the theory of knots. Volume 1 in the series is Louis H Kauffman's Knots and Physics.

One purpose of this series is to continue the exploration of many of the themes indicated in Volume 1. These themes reach out beyond knot theory into physics, mathematics, logic, linguistics, philosophy, biology and practical experience. All of these outreaches have relations with knot theory when knot theory is regarded as a pivot or meeting place for apparently separate ideas. Knots act as such a pivotal place. We do not fully understand why this is so. The series represents stages in the exploration of this nexus.

Details of the titles in this series to date give a picture of the enterprise.

More information on this series can also be found at http://www.worldscientific.com/series/skae

K&E Series on Knots and Everything — Vol. 76

Combinatorial Knot Theory

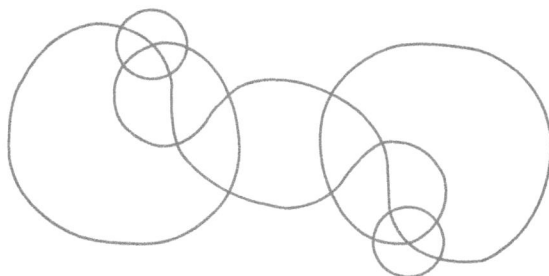

Roger A Fenn

University of Sussex, UK

World Scientific

NEW JERSEY · LONDON · SINGAPORE · BEIJING · SHANGHAI · HONG KONG · TAIPEI · CHENNAI

Published by

World Scientific Publishing Co. Pte. Ltd.
5 Toh Tuck Link, Singapore 596224
USA office: 27 Warren Street, Suite 401-402, Hackensack, NJ 07601
UK office: 57 Shelton Street, Covent Garden, London WC2H 9HE

Library of Congress Cataloging-in-Publication Data
Names: Fenn, Roger, 1942–   author.
Title: Combinatorial knot theory / Roger A. Fenn, University of Sussex, UK.
Description: New Jersey : World Scientific, [2025] |
    Series: Series on knots and everything, 0219-9769 ; vol. 76 |
    Includes bibliographical references and index.
Identifiers: LCCN 2024033095 | ISBN 9789811292729 (hardcover) |
    ISBN 9789811292736 (ebook for institutions) | ISBN 9789811292743 (ebook for individuals)
Subjects: LCSH: Knot theory. | Combinatorial geometry.
Classification: LCC QA612.2 .F46 2025 | DDC 514/.2242--dc23/eng/20240905
LC record available at https://lccn.loc.gov/2024033095

British Library Cataloguing-in-Publication Data
A catalogue record for this book is available from the British Library.

For any available supplementary material, please visit
https://www.worldscientific.com/worldscibooks/10.1142/13826#t=suppl

Desk Editors: Kannan Krishnan/Angeline Husni

Typeset by Stallion Press
Email: enquiries@stallionpress.com

Roger Fenn: A Life in Maths

Roger Fenn, with his keen interest in maths, always wondered about recurring decimals, extracting square roots, and the beautiful Greek symbols in his older brother's school books. However, this didn't prevent him from failing the local 11-plus exam which was universal at the time. The passers were sent to prestigious establishments such as the Bristol Grammar School and the losers went to the Filton Road Secondary Modern School, where to his surprise he was in the top stream and began to learn proper maths and even French and German.

Filton Road only survived another year and the pupils were moved up the hill to Lockleaze School, the first comprehensive school in Britain. In evidence that if you spend money you get results, Lockleaze had a large playing field, a physics lab, a woodwork room, and a metalwork room with a kiln. But most importantly, it had good teachers, a head of music, a head of languages, and a head of maths, Colin Evans. It was Colin who taught Roger all he knew and got him into Clare College, Cambridge, in 1961, where he was the first 11-plus failure and comprehensive pupil to attend. After Cambridge, he did his PhD under John Reeve at King's College, London, where he learnt about geometric topology and proved 'the table theorem' for his thesis, which showed one could always place a square table so that it didn't wobble and the top was horizontal.

In 1967, Roger became a lecturer at the new (plate glass) University of Sussex, where he stayed for the rest of his career. According to the Research Gate, he has 93 publications to his credit, with more to come. Roger is now retired from teaching but still doing maths. He lives in Lewes, Sussex, with Liz his wife and constant companion and enjoys walking, making things, and playing music badly.

Combinatorial Knot Theory:
An Introduction to Knot Diagrams

Combinatorial Adj.(2) *of or relating to the arrangement of, operation on, and selection of discrete mathematical elements belonging to finite sets or making up geometric configurations.*
— Merriam-Webster

Knot theory, like ancient Gaul, can be divided into three. There is the traditional or topological view of a knot as a flexible curve tied in space. This object can then be projected onto a plane as a **diagram** and information deduced from this. Finally, algebraic information can be deduced from both views. In this book, we will consider the second of these. Combinatorial knot theory is the study of knots from the point of view of a diagram and how it changes under the **Reidemeister moves** [235]. These two concepts form the basis of this book and are explained in the following.

Traditionally, a **knot** is considered as a loop or closed curve in 3-dimensional space with three coordinates x, y, z. It is convenient to consider the x, y plane as horizontal and the z coordinate as height. The knot can then be represented by a projection onto the x, y plane. If the curve is smooth or polygonal then by a small adjustment of the curve in space, the image of the knot can be a smooth or a polygonal curve in the plane with a finite number of **crossings** or

double points. Above a crossing point in the horizontal plane are two points in space, an **overcrossing** point and an **undercrossing** point. Another description of the image is as a 4-valent planar graph or network in the horizontal plane with the vertices corresponding to these double points. At this stage, we have the **shadow** of the knot defined by the projection. We can reconstruct the knot by lifting the vertices of the graph into two points in space, one above the other.

We can represent the relation between the knot and its image in the plane by breaking the arc containing the undercrossing point so that an observer above the knot would see the overcrossing point obscuring the undercrossing point. This break converts the image into a **diagram** which is the central concept of the whole book. From a diagram, you can convert this 2-dimensional object back to the original knotted curve.

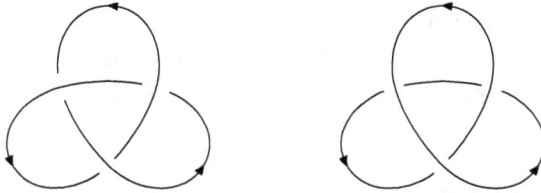

Fig. 1. Two knot diagrams.

In Fig. 1, the projected shadows are identical, but the diagrams are different as are the knots above them. The left-hand knot is the non-trivial trefoil and the right-hand knot is equivalent to the unknot or flat circle. Since S^1 is oriented, say anti-clockwise, it follows that the image curves are oriented and this is indicated by the little arrows.

We now come to the problem of how the diagram changes when the curve in space changes because we are only concerned with the topological properties of the knot. It turns out that we can think of the diagram changes as the product of a finite number of moves named after the German mathematician, Kurt Reidemeister (1893–1971). So, if we have some property, invariant under the Reidemeister moves, then this property will be a topological property of the knot itself.

The moves themselves are now illustrated in Figs. 2–4. They are local in character in the sense that outside the three illustrations, nothing changes.

The first Reidemeister move removes or introduces a single loop.

Fig. 2.   The first Reidemeister move.

The second Reidemeister move moves an arc which defines the side of a 2-cornered oval either over or under the other side. There are two arcs, one above the other.

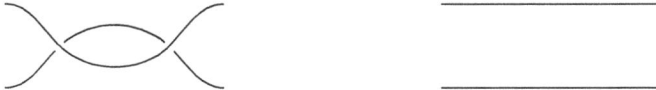

Fig. 3.   The second Reidemeister move.

The third Reidemeister move changes the sides of a 3-cornered region. There are three arcs: one on top, one on the bottom and one in the middle.

Fig. 4.   The third Reidemeister move.

There now follows the reader's first exercise. These vary in difficulty, but you could always look for inspiration in Wikipedia (other search engines are available).

Exercise 1. *In Fig. 4, illustrating the third Reidemeister move, it seems as though it is the middle arc that moves. Convince yourself that topologically it could also be the top or the bottom arc that moves.*

Exercise 2. *Tie some string to copy the right-hand knot in Fig. 1. Show that the right-hand example can be undone by a Reidemeister 2 move followed by a Reidemeister 1 move.*

Exercise 3. *Tie some string to copy the left-hand knot in Fig. 1. Show that the orientation of the left-hand example indicated by the arrows is irrelevant. This example is called the* **right**-**hand trefoil.**

Exercise 4. *Now, tie some string to mimic the mirror image of the left-hand example to make the* **left**-**hand trefoil** *and convince yourself that the two variants are different.*

In this book, we will look at generalisations of diagrams and the Reidemeister moves. This will be explained in the text. However, a brief description can be given here. The optics at the crossing points of a classical knot diagram indicate how the knot is embedded in space, whether one small piece of arc is above or below the other. This is called a *glyph*. The overcrossing and undercrossing points of the classical glyph could be replaced by various glyphs. For example, a circle containing the crossing is used for a virtual crossing, but since this also indicates a weld crossing, a more exact notation is needed. This is achieved by a label or *tag*. The tags are lowercase letters from the Roman alphabet, such as r for classical (real), v for virtual and w for weld. This is the stuff of combinatorial knot theory and its generalisations which form the contents of this book.

As is the way with mathematics, which makes it simultaneously infuriating and appealing, combinatorial knot theory and the topology of knots are two sides of the same coin and yet the connection between some aspects of the two has not been and may never be discovered. For example, the crookedness, stick number and energy are knot invariants which as far as I know have not been defined from their diagrams. On the other hand, the fundamental biquandle and the crossing number are only definable from a diagram: again, as far as I know. Hopefully, this book may inspire future researchers to find some pathways through the thickets. We will not define these invariants in this volume, but a search of the internet will provide answers, see the exercise.

Exercise 5. *In the above paragraph are five knot concepts you may not have met. See if you can find their definitions on the web.*

This work and the book would not have been possible without the brilliant collaboration and insights from the following mathematicians who have helped me over the decades with my stuttering research. They are Andy Bartholomew, Steve Budden, Mercedes Jordan, Seiichi and Naoko Kamada, Lou Kauffman, Vassily Manturov, Richard Rimanyi, Dale Rolfsen, Colin Rourke, Brian Sanderson, and Vladimir Turaev. Anyone feeling aggrieved by being left out should contact me.

There have been various books and papers published on combinatorial knot theory. Initial seminal papers were by Alexander [1] and Reidemeister [235]. Books by Kauffman [107] and Manturov [189] together with other important papers are included in the bibliography.

For convenience of referring back, figures, sections, theorems, etc. in this book are indicated by three numbers. In order they are, the chapter number, the section number and the count within the section.

Contents

Chapter 1

Basic Concepts

To solve math problems, you need to know the basic mathematics
before you can start applying it.
— Catherine Asaro

1.1 Introduction

This book is meant to be fairly self-contained so that anyone who has studied some higher mathematics should be able to follow the arguments from cover to cover. This is probably totally utopian on my behalf, for which I apologise. I know that I have done this journey from front to back and so I hope that others can. But if you get something from this book, then I will be happy.

We start off with the very basics. A general knot theory is defined by diagrams and how they change under Reidemeister or R **moves**. So, the first job is to define these and related concepts. In order not to interrupt the flow of the narrative, we will be fairly cavalier in our definitions and often appeal to intuition. Concerned readers can read any number of books which will give an appropriate hard core definition. For example, there are fine books on general topology by Dugundji, *Topology* (Allyn and Bacon, 1966) or *General Topology* by Kelley, although, oddly enough, a proper definition seems remotely different from an informal one. One can read these two books and

wonder what on Earth they have to do with knot theory! I have tried
to make this book accessible to any student who has done a course in
elementary linear algebra and calculus. In this book, I will use certain
technical terms like "isotopy", "homotopy", "homeomorphism", etc.
The real meaning of these words is easy to grasp in a hand-wavy way
and I make no apologies for giving such a definition as well as a more
technical one. I urge the student who comes to these concepts for the
first time to try and understand these definitions informally. After
all, we are only talking about knotted strings in space and curves
drawn in the plane. Everyone, not just mathematicians, can have an
intuitive handle on those two ideas.

1.2   The nouns and verbs of mathematics

Like any other language, mathematics has its nouns and verbs.
Loosely speaking, the nouns of mathematics define concepts which
are fixed while the verbs move them around. However, like English,
these ideas can be mixed up.[1] We start off by considering objects
and notation, and then a discussion of the new concepts and definitions needed to describe the mathematical ideas, the nouns of
mathematics. Finally, we look at words which do things to our newly
defined objects, the verbs of mathematics. More seasoned mathematicians can jump straight to Section 1.6.

1.2.1   *Objects and notation*

We start off with a list of concepts defined by words which are the
noun objects and immediately fly into a problem. Is a function a
noun or a verb? Since it implies movement, it ought to be a verb,
but its fundamental nature means it has to be defined early on.

   Many of the concepts are defined by words which may be new or
new in their usage to the reader. Mathematicians love inventing new
words and new meanings for old words. Topologists are no exception.

---

[1]Metaphysically speaking, we could think of mathematics as a vast category with
objects and morphisms (see Appendix B).

The **real numbers**, denoted by $\mathbb{R}$, can be considered as a line. The subset of **integers** or whole numbers are denoted by $\mathbb{Z}$. The **positive integers**, $1, 2, 3, \ldots$ are denoted by $\mathbb{N}$.

Exercise 1.1. *Draw a line with a ruler and mark all the integers from -1 to 5. Put in where you think $\pi, e, \sqrt{2}$ and the golden ratio should go.*

The cartesian product of n lines, $\mathbb{R}^n$, is the **n-dimensional real space**. A typical point in $\mathbb{R}^n$ has n coordinates and can be written as $\mathbf{x} = (x_1, x_2, \ldots, x_n)$, where $\{x_1, x_2, \ldots, x_n\}$ are arbitrary elements of $\mathbb{R}$.

Another type of space is a **topological space**. All of our topological spaces are subsets of $\mathbb{R}^n$, where $n \leq 4$, and are therefore **metric spaces**. The distance metric is $d(\mathbf{x}, \mathbf{y}) = \sqrt{(x_1 - y_1)^2 + \cdots + (x_n - y_n)^2}$, which is a generalisation of the famous Pythagorean formula if $n = 2$. The standard **circle**, S^1, is the set of points in the x, y plane, distance 1 from the origin. So, $S^1 = \{(x, y) \in \mathbb{R}^2 \mid x^2 + y^2 = 1\}$. This is a particular example of the **n-sphere**, S^n, given by the equation $x_1^2 + x_2^2 + \cdots x_{n+1}^2 = 1$ where $(x_1, x_2, \ldots x_{n+1})$ lies in $\mathbb{R}^{n+1}$. The 2-sphere S^2 is a **surface** of **genus** 0.

A closed connected compact oriented surface, Σ, of genus g can be described as follows. Think of the sphere S^2 as a horizontal plane with a point at infinity. Take a pair of disjoint disks in the plane and remove the interiors. Glue the two boundaries at b to make a handle, as shown in Fig. 1.1. Do this g times.

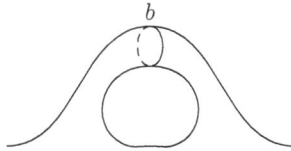

Fig. 1.1. Raising a handle.

Let Σ_g^r denote a surface of genus g with the interiors of r disjoint disks removed. Any connected, compact, oriented surface is of this form for some g and r, see the work of Moise [222].

Continuity: One definition of topology is the study of continuity. We can take a ball of dough and shape it, pull it, and pummel it without breaking it. These are all continuous actions. On the other hand, if we cut it in two, then we have performed a discontinuous action. Technically, we say that a function is **continuous** if it preserves convergent sequences. Let $(x_n), n \in \mathbb{N}$ be a sequence of points in a space X which converges to a point x. The function $f : X \to Y$ is **continuous** if for all such sequences, $f(x_n)$ converges to $f(x)$. There are more general definitions of a continuous function, but this will be a suitable definition for our purposes.

Exercise 1.2. *Show that the function $f : \mathbb{R} \to \mathbb{R}$ defined by*

$$f(x) = \begin{cases} 1 & \text{if } x > 0 \\ -1 & \text{otherwise} \end{cases}$$

is not continuous by finding a sequence (x_n) converging to some point x, with an image sequence, $f(x_n)$, which doesn't converge to $f(x)$.

Smoothness: If we draw a continuous path in the plane, we can easily imagine it as **smooth** or perhaps not smooth if it has a kink in it. Technically, the actual definition is more difficult to formulate. Let a curve in space be defined by three functions, $x = f(t)$, $y = g(t), z = h(t)$ of a parameter t, say. Then this curve is said to be **smooth** if the three functions can be infinitely differentiated and the tangent $(\dot{f}, \dot{g}, \dot{h})$ always has positive length. The notation $\dot{f}$ is due to Newton and means $\frac{df}{dt}$, which was Liebnitz' notation.

Exercise 1.3. *Show that the curve in space given by $\mathbf{x}(t) = (t, t^2, t^3)$ is smooth for all values of t, but that its projection in the y, z plane is not smooth when $t = 0$.*

A smooth map is called an **immersion**. We would like our immersions to be in **general position**. Informally, this means that the intersections where it fails to be an embedding are as small as possible. So, if a space of dimension i and a space of dimension j are immersed in a space of dimension n in general position, then the intersections between the two spaces should have dimension not greater than $i + j - n$ and should be stable in some sense. So, a curve in 3-dimensional space would in general position be an embedding, but a general position curve in a surface would meet only in transverse double points where the tangents cross. Luckily, we only have these very low dimensions to consider. We don't have to worry about a general position surface in 3 dimensions with difficult to visualise self-intersections consisting of double lines, triple points and branch points.

Informally, a **homeomorphism** is a function that preserves the topological properties of a space. So, geometrical properties and spatial relations unaffected by the continuous change of shape (or size) of figures are preserved. For example, a circle and the boundary of a square are related by a homeomorphism, i.e. they are homeomorphic. More formally, a homeomorphism is a continuous bijection with a continuous inverse.

Exercise 1.4. *Let $h : [0, 1) \to S^1$ from the half-open interval to the circle be defined by $h(x) = e^{2\pi i x}$. Show that h is a continuous bijection but not a homeomorphism.*

Exercise 1.5. *Find homeomorphisms between the square, $\{(x, y) \in \mathbb{R}^2 \mid \max(|x|, |y|) \leq 1\}$, the diamond, $\{(x, y) \in \mathbb{R}^2 \mid |x| + |y| \leq 1\}$ and the circular disk, $\{(x, y) \in \mathbb{R}^2 \mid x^2 + y^2 \leq 1\}$. Deduce that the boundaries of a square and a diamond are homeomorphic to S^1.*

An **embedding**, $X \to Y$, is a homeomorphism from X to a subspace of Y. So, a classical knot is represented by an embedding of a circle, S^1, into space, $\mathbb{R}^3$ or the 3-sphere, S^3. The knot is **tame** if the embedding is smooth or can be made smooth in some continuous sense. Otherwise, the knot is **wild**. All the knots in this book will be tame.

1.2.2 *Movement*

If the previous sections were loosely about the *nouns* of mathematics, then this next section is about the *verbs* of mathematics, topological words which imply motion and so are dependant on a variable which could be time. **Homotopy** is the vaguest way of recognising various topological properties. The idea is that you take a function and vary it continuously. Let $f, g : X \to Y$ be two functions.[2] Then we say that they are **homotopic** if there is a function $F : X \times I \to Y$, where $I = [0, 1]$ is the closed unit interval and $F(x, 0) = f(x)$ and $F(x, 1) = f(y)$. We can think of this as a path between the functions in a function space. In fact, if $X = \{x\}$, a single point, then a homotopy is a path in Y.

An **isotopy** is a homotopy of homeomorphisms or embeddings. An **ambient isotopy** is an isotopy of a space containing a subspace. If two immersions in general position are isotopic, then during the course of this isotopy, there may be values of the parameters when the immersion is not in general position. We call these exceptional parameters **singular values**. But the methods of general position may also be applied to isotopies so that these singular values will be as small as possible and fall into a well-defined set of objects. In general, these general position isotopies are very complicated and mathematical hours have been spent studying them. Luckily, because the dimensions we study are very low, we can easily guess what form they take and we will consider our examples later in this chapter.

In the same way that the above definitions are easily visualised due to the low dimensions we are working with, the concept of **orientation**, which normally involves the use of homology theory, can also be defined. Points are not normally oriented, but they can be labelled or tagged as we shall see. Curves are oriented by an arrow which implies some sort of forward motion. Surfaces are oriented by small circles inheriting a clockwise or anticlockwise direction. Three-dimensional manifolds are oriented by small oriented spheres and normals satisfying the right-hand rule.

Exercise 1.6. *A super-intelligent hyper-being from alpha centauri phones you up and asks you to explain over the telephone the right-hand rule for orienting space. How do you respond? Repeat for the offside rule in football.*

Exercise 1.7. *What direction is widdershins, sunward?*

---

[2]Functions are sometimes called maps for obvious reasons.

1.3 Shadow diagrams

This completes the lexicon needed for elementary topology. All future definitions will be related to the book and be given in the text. For the purposes of this book, a **curve** in a surface Σ is a smooth function $\alpha : \bigsqcup S_i^1 \to \Sigma$ in general position where $\bigsqcup S_i^1$ is a set of disjoint circles. We will often call these **shadow diagrams** for reasons which will become clear. The restriction of α to any S_i^1 is called a **component curve**.[3]

Curves, $\alpha_1 : \bigsqcup S_i^1 \to \Sigma_1$ and $\alpha_2 : \bigsqcup S_j^1 \to \Sigma_2$, are said to be **topologically equivalent** if there are homeomorphisms, $h : \bigsqcup S_i^1 \to \bigsqcup S_j^1$ and $k : \Sigma_1 \to \Sigma_2$, which preserve orientation, such that the following diagram commutes:

$$
\begin{array}{ccc}
\bigsqcup S_i^1 & \xrightarrow{\alpha_1} & \Sigma_1 \\
\downarrow h & & k \downarrow \\
\bigsqcup S_j^1 & \xrightarrow{\alpha_2} & \Sigma_2
\end{array}
$$

so $k \circ \alpha_1 = \alpha_2 \circ h$.

The portions of the curve between the crossings points are called **edges**. The connected areas of the surface not on the image of the curve are called **regions**.

The following figure gives two examples of planar shadow diagrams: the "trefoil" with 1 component, three crossing points, six edges and five regions consisting of two trigons, one of which is unbounded, and three bigons, and the "Borromean rings" with three components, six crossing points, 12 edges and eight regions, all of which are trigons. Note that in both cases the alternating sum has value $3 - 6 + 5 = 6 - 12 + 8 = 2$, and this is general.

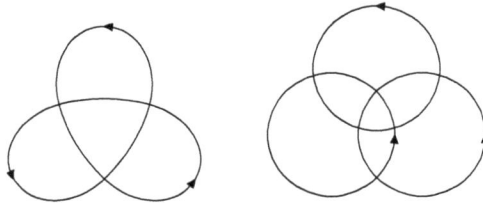

Fig. 1.2. Shadow diagrams.

---

[3]We don't care if the number of components is more than one. We still call it a curve. A similar elision is used for knots and links.

These diagrams are "shadows" in the sense that are both images or shadows in the plane of projected curves in space.

Exercise 1.8. *Show that the trefoil shadow diagram in Fig. 1.2 is the image of three distinct knots in space (you may assume that there are two trefoils a right and a left).*

Exercise 1.9. *How many distinct knots project onto the Borromean rings in Fig. 1.2?*

Exercise 1.10. *Show that a regular curve in the plane with n crossings has $2n$ edges and the number of regions is $n+2$, one of which is unbounded. Hint: for this exercise and the next, use Euler's formula that $\#\{vertices\} - \#\{edges\} + \#\{bounded\ regions\} = 1$.*

Exercise 1.11. *Show that any curve in the plane or sphere with connected image and with at least one crossing has either a monogon, bigon or trigon region. Moreover, if there are no monogons or bigons, then there are at least eight trigons.*

Consider now an oriented graph on a surface with vertices, $V_1, \ldots, V_n$ and edges, $E_1, \ldots, E_m$, none of which are doubled, i.e. have the same end points. Each edge is oriented from an initial vertex, V_i, to a final vertex, V_j, and we sometimes write this edge as $V_i V_j$. If the graph is the image of a curve in general position, then the valency of a vertex, V, is four, with two incoming edges which have V as their final vertex and two outgoing edges which have V as their initial vertex. Moreover, around V, the incoming edges are adjacent as are the outgoing edges, like local x and y coordinate axes.

Exercise 1.12. *Let Γ be an oriented graph on a surface. Suppose that at each vertex, V, of Γ, there are two edges E_W and E_S which have V as their final vertex and two edges E_E and E_N which have V as their initial vertex. If these edges have the cycled order $E_E, E_N, E_W,$ and E_S around V, show that Γ is the image of a curve.*

1.4 Knot diagrams[4]

We now associate curves and their attachments with a generalised knot. By the usual abuse of notation, this augmented curve is often confused with the knot it represents.

Knot diagrams, or diagrams for short, consist of the following:

(1) An immersion in general position of a finite number of circles in an oriented surface, Σ, i.e. a curve in Σ: if the surface is a plane or sphere, it is called a **planar diagram**. The image of one of the circles is called a **component** of the diagram.

(2) The double point crossings fall into **types** tagged by a roman letter, a, say. The tags lie in a tag-bag, $\mathcal{T}$. There is an involution, $\mathcal{T} \to \mathcal{T}$ notated, $a \to \bar{a}$, and called the **inverse**. Tags for which $a \neq \bar{a}$ are called **3-dimensional or spacial tags**. The tags which are not 3-dimensional are called **2-dimensional or flat tags** and so $\bar{a} = a$ in this case. Since the map is an involution, in all cases, the inverse of $\bar{a}$ is a, and so, $\bar{\bar{a}} = a$.

(3) Two diagrams, K and L, are considered the same if there is a homeomorphism of Σ which takes one immersion to the other and preserves orientation and tags. In that case, we write $K \cong L$ and call them **isomorphic**.

Diagrams are usually denoted by, $K, L, M, \ldots$ The **shadow** of a diagram K is denoted by $|K|$ and is the curve of K with all the tags ripped off. Figure 1.3 is an example of a diagram.

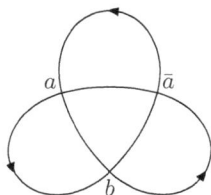

Fig. 1.3. Diagram with shadow the trefoil curve and tags $a, \bar{a}, b$.

---

[4]The definition of knots also includes links.

The empty tag, o, is also a tag and is 2-dimensional, so $o = \bar{o}$. A diagram, K, with all empty tags satisfies $|K| = K$.

Sometimes, the tag is incorporated into the crossing itself in the form of a **glyph** at the crossing. Classical crossings with tag r for (r)eal are indicated by a glyph which is a break in the curve. In Fig. 1.4, the two types of classical crossings are illustrated. The crossing on the left with tag $\bar{r}$ is a **left-hand** or **negative** crossing. The crossing on the right with tag r is a **right-hand** or **positive** crossing.

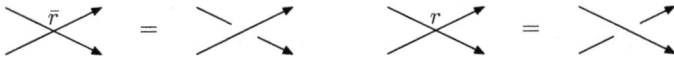

Fig. 1.4.   Tags and glyphs on negative and positive real crossings.

The real tag r is 3-dimensional.

We shall introduce more tags and glyphs later. In the meantime, the Roman letters $a, b, c, \ldots, x, y, z$, usually small, will stand for general tags, but some letters like r and o are reserved for particular tags.

1.5   Generalised Reidemeister or R moves

To continue with our definition of a knot theory, we take a first look at generalisations of the famous **Reidemeister moves**. These are designated R_1, R_2, and R_3 with order corresponding to the original order. These moves change one diagram into another. In such a move, a neighbourhood of a small disk portion of the diagram is changed, but the rest of the diagram stays fixed.

1.5.1   *Simple knot theories*

Simple knot theories only involve one type of crossing designated by the tags a and $\bar{a}$. If a is 2-dimensional, then $\bar{a} = a$, otherwise $\bar{a} \neq a$. An example of a simple theory is classic knot theory. The tags on classic knots are r and $\bar{r}$, but historically, they are represented by their glyphs. We will discuss which R moves are allowed in this case later. The reader may care to replace the crossings below by the glyph for the real crossing r to see that they coincide with the

Reidemeister moves shown earlier. Shadow diagrams form a trivial simple theory in which no R moves are allowed.

R_1, **the first R move:** This move may be an expansion or a retraction, either of which is the inverse of the other. In an R_1 expansion, a monogon is created (shaded in the figure) and the resulting crossing is tagged. We write $R_1(a)$ if the crossing is tagged with a. In Fig. 1.5, the movement from right to left is an expansion with tag a. The movement from left to right is a retraction with tag a. We write $\bar{R}_1(a)$ for this motion. Needless to say, in a retraction, no part of the diagram can lie in the interior of the shaded portion and the same applies to all the following moves.

Fig. 1.5. $R_1(a)$: an R_1 move with tag a.

If orientation of the arc is taken into account, then the shaded monogon may be oriented clockwise or anticlockwise.

R_2, **the second R move:** This move may also be an expansion or a retraction, and again, the two are inverse to each other. In an R_2 expansion, a shaded bigon is created and the resulting crossings are tagged by an inverse pair. In Fig. 1.6, the movement from right to left is an expansion with tags a and $\bar{a}$.[5] We write $R_2(a)$ for this and $\bar{R}_2(a)$ for the movement from left to right.

Fig. 1.6. $R_2(a)$: A parallel R_2 move with tags a and $\bar{a}$.

In Fig. 1.6, the orientation for both arcs is from left to right. We call this a **parallel R_2 move**. In Fig. 1.7, the orientations of the arcs are opposite. This R_2 move is called a **non-parallel R_2 move**.

---

[5]The tag a may be replaced by $\bar{a}$ and it makes sense that $\bar{\bar{a}} = a$.

This second type in which the orientations of the arcs are opposite has an important submove called a Vogel move. We will consider this in a later chapter. In the meantime, we will include both moves under R_2.

Fig. 1.7. $R_2(a)$: a non-parallel $R_2(a)$ move with tags a and $\bar{a}$.

Knot theories which allow $R_2(a)$ for all tags a are called **regular**. The only knot theory which does not allow R_2 or indeed any R moves is the shadow theory.

R_3, the third R move: It will come as no surprise to know that this move involves a trigon. The third Reidemeister move transforms the central trigon into another trigon. It can involve up to three tags and the general R_3 move is denoted by $R_3(a, b, c)$. This move is illustrated in Fig. 1.8.

Fig. 1.8. An $R_3(a, b, c)$ move.

Note the orientation from left to right. The other possible orientation orients the central trigon either clockwise or anticlockwise. But for regular theories, we will see later that this move can be transformed into the one above.

The middle arc starts with the tags a and c in order from left to right. After the move, the tags in the middle arc are reversed. This means that the anticlockwise orientation abc is preserved under a rotation through 120 degrees. The crossing tagged by b moves from bottom to top.

Exercise 1.13. *Show that for a regular theory, the six moves,* $R_3(ab, c)$, $R_3(b, a, \bar{c})$, $R_3(\bar{b}, \bar{c}, a)$, $R_3(\bar{c}, \bar{b}, \bar{a})$, $R_3(c, \bar{a}, \bar{b})$, *and* $R_3(\bar{a}, c, b)$, *are equivalent.*

The rule for the above exercise is: swap the two consecutive tags either on the left or right and bar the other tag.

Exercise 1.14. *Use the above exercise to show that the moves,* $R_3(a, a, b)$, $R_3(a, a, \bar{b})$, $R_3(\bar{a}, \bar{b}, a)$, $R_3(\bar{b}, \bar{a}, \bar{a})$, $R_3(b, \bar{a}, \bar{a})$, *and* $R_3(\bar{a}, b, a)$, *are equivalent.*

Exercise 1.15. *Show that if the tags in Fig. 1.8. are real, say,* $a = r, c = \bar{r}$, *then the arc which starts in the middle appears to move over the crossing tagged b.*

Exercise 1.16. *Show that the* R_3 *moves,* $R_3(r, r, r)$, $R_3(r, r, \bar{r})$, $R_3(r, \bar{r}, \bar{r})$, $R_3(\bar{r}, \bar{r}, \bar{r})$, $R_3(\bar{r}, r, r)$, $R_3(\bar{r}, \bar{r}, r)$, *are plausible in three dimensions, but* $R_3(r, \bar{r}, r)$, *and* $R_3(\bar{r}, r, \bar{r})$ *are implausible.*

1.6 Compound knot theories

A **compound knot theory** uses more than one type of tag, although in practice, only two are ever used. Later, we will introduce a fourth R move, R_4. In the meantime, we will consider the problem of orientation for R_3.

1.6.1 *Orienting the arcs of the R_3 move*

Orienting the R_2 move involved two distinct orientations. It turns out that we need only one orientation of the R_3 move, the parallel one where the orientations are all in one direction, see Fig. 1.8. This is achieved by changing views of the various cases, a trick due to Turaev and judicious choices of tags. Readers who are not interested in the details of the proof can accept that fact and move to the following section.

The remaining readers should now consult Fig. 1.9. which for convenience is drawn with straight arcs and is a representative of the most general R_3 move. Moreover, we have rotated the diagram so that the middle line is oriented from left to right. Furthermore, we assume that the central trigon is an equilateral triangle.

The symbols u and v attached at the left-hand side of the two diagonal arcs indicate possible orientation and take the values $\pm$. So, if $u = +$, then the arc is oriented from the top left to the bottom right. If $u = -$, then the opposite is true. Similarly, if $v = +$, then the arc is oriented from the bottom left to the top right.

On the left of the diagram, the middle oriented line passing through the crossings of type, a and c, changes to an oriented line passing through the crossings of type, c and a, and appears to move from the top to the bottom, the crossing types being interchanged in order, passing b. However, this is only an illusion. We can also think of the NW/SE line passing through the crossing labelled c or the SW/NE line passing through the crossing labelled a and in each case reversing the tags.

We indicate this general move from left to right by the symbol, $R_3(u, v : a, b, c)$ and its reverse by $\bar{R}_3(u, v : a, b, c)$.

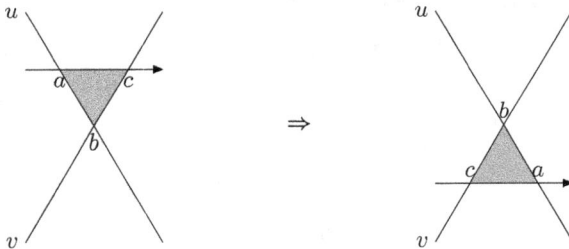

Fig. 1.9. $R_3(u, v; a, b, c)$, the most general R_3 move.

At first sight, it seems there are at least $2 \times 2 \times 6 = 24$ possibilities for the moves, but these can be reduced by the following sequence of lemmas.

Let $R_3(+, + : a, b, c)$ be denoted by $R_3(a, b, c)$ and write

$$R_3(u, v : a, b, c) \leftrightarrow R_3(u', v' : a', b', c')$$

to mean if $R_3(u, v : a, b, c)$ is allowed, then this implies that $R_3(u', v' : a', b', c')$ is allowed and conversely.

Write

$$R_3(u, v : a, b, c) \to R_3(u', v' : a', b', c')$$

to mean if $R_3(u, v : a, b, c)$ is allowed, then this implies $R_3(u', v' : a', b', c')$ is allowed, but not necessarily conversely.

Lemma 1.1. $R_3(a, b, c) \leftrightarrow R_3(-, +; b, c, a) \leftrightarrow R_3(+, -; c, a, b)$.

Proof. First rotate Fig. 1.9. with $u = v = +$ through $\pi/3$ clockwise and then $\pi/3$ anticlockwise. This results in the inverses of the last two R_3 moves being implied. □

We call the three moves in the above lemma **parallel** because the arcs can always be oriented so that they appear to point in the same direction.

Lemma 1.2. $R_3(-, -; a, b, c) \leftrightarrow R_3(-, -; c, a, b) \leftrightarrow R_3(-, -; b, c, a)$.

Proof. First rotate Fig. 1.9. with $u = v = -$ through $2\pi/3$ clockwise and then $2\pi/3$ anticlockwise. □

The three moves in the above lemma have oriented trigons and are called **non-parallel**.

The following lemma uses a **trick of Turaev** to convert an unknown R_3 move to a sequence of R_2 move, known R_3 move and R_2 move. So, we assume that the theory is regular. This is no hardship as the only non-regular theory is the shadow theory and it has no R_3 moves anyway.

Fig. 1.10. $R_3(u, v; a, b, c) \to R_3(v, u; \bar{c}, b, a)$ Turaev's trick.

In Fig. 1.10, we assume that $R_3(u, v; a, b, c)$ is allowed and is about to start, but a creative $R_2(c)$ move has occurred. When the $R_3(u, v; a, b, c)$ move has taken place, a retractive $R_2(c)$ move implies that an $R_3(v, u; \bar{c}, b, a)$ move is completed.

The following lemma is implied by the above results.

Lemma 1.3. $R_3(a, b, c) \leftrightarrow R_3(b, a, \bar{c})$, $R_3(a, b, c) \rightarrow R_3(-, -; \bar{b}, c, a)$.

$\square$

Summing up, we see that we only need to specify the $R_3(a, b, c)$ move.

1.7 Dominance and the fourth R move

As mentioned earlier, compound knot theories only have two types of tags in practise. The individual tags satisfy their allowed $R(a)$ moves discussed earlier and the combination R move takes two forms. We say that the tag x **dominates** the tag y if $R_3(x, x, y)$ is allowed. We write $x \succ y$ for this relation. Using the above lemmas, we have the following.

Theorem 1.1. *The following statements are equivalent:*

(1) *x dominates y;*
(2) *$R_3(x, x, y)$ is allowed;*
(3) *$R_3(x, x, \bar{y})$ is allowed;*
(4) *$R_3(\bar{x}, y, x)$ is allowed;*
(5) *$R_3(y, \bar{x}, \bar{x})$ is allowed;*
(6) *$R_3(\bar{x}, \bar{y}, x)$ is allowed;*
(7) *$R_3(\bar{y}, \bar{x}, \bar{x})$ is allowed.*

We deduce from Theorem 1.1 that if a tag x dominates a tag y, then it also dominates the inverse tag $\bar{y}$. However, in that situation, we cannot deduce that $\bar{x}$ dominates y. $\square$

If there is an x such that x and $\bar{x}$ dominate every tag in the theory, we say that x dominates the theory and call the theory **normal**.

Fig. 1.11. x dominating a (see (4) in Theorem 1.1).

Classical knot theory is normal. Both r and $\bar{r}$ dominate. It is easily seen that r dominates r and $\bar{r}$, and similarly $\bar{r}$ dominates r and $\bar{r}$.

In a classical knot theory diagram, it is possible for a subpath of the diagram to lie above the rest of the diagram and so, keeping its end points fixed, this subpath can be moved, changing the diagram. Something similar occurs for subpaths lying below the diagram. We can make this idea for a general normal theory precise as follows.

Suppose a tag x dominates a normal theory. A subpath, P, of a component of a diagram is said to be (x) **above** if the only crossings it meets are one or more of the two types illustrated on the left of Fig. 1.12. The portion of the subpath, P, illustrated is drawn with a thicker line.

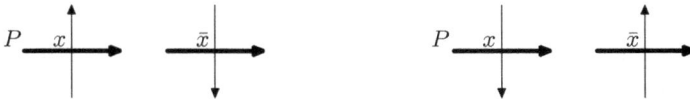

Fig. 1.12. x above and x below paths.

Similarly, P is said to be (x) **below** if the only crossings it meets are the ones on the right of Fig. 1.12. If x is r, the classical tag, then above and below have their usual meanings.

1.7.1 *The virtual tag*

This is a good moment to introduce a very important tag and concept. The **virtual tag**, v, satisfies all the single $R(v)$ moves defined above with one important extra property; v dominates all other tags. The glyph for a virtual crossing is a flat crossing enclosed by a small circle (see Figure 1.13).

Fig. 1.13. Virtual crossing.

The **detour** move, as shown by Kauffman [117], is defined by the following lemma.

Lemma 1.4 (The detour move). *Let P be an x above/below sub-path of a diagram K in a normal theory and let P' be a path with the same end points as P which crosses K in such a manner as to create an x above/below path. Then the diagrams K and $(K - P) \cup P'$ are related by a sequence of R moves.*

Proof. By a simplification of where the paths cross, we can assume that $P \cup P'$ is the boundary of a bigon and argue by induction on the number of crossings inside. If there are none, then an arc entering the bigon either leaves from the same K with all the tags ripped off.side or the opposite side. If the former, then an innermost arc can be eliminated by an R_2 move. Eventually, all the arcs cross from one side to the other. Then P and P' are isotopic.

If the bigon contains crossings, then they can be eliminated one by one using finger moves in which a small part of P or P' moves to enclose the crossing.   □

To illustrate the detour move, consider Fig. 1.14. The example, due to Kauffman [117], has two real and two virtual crossings.

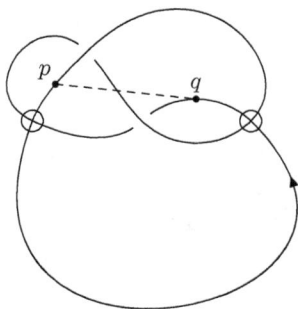

Fig. 1.14.   Kauffman's example.

The subpath oriented from p to q passes through two virtual crossings, and so, we can move it to the dotted line from p to q. After smoothing, we get the diagram in Fig. 1.15, which has one less virtual crossing.

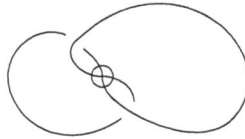

Fig. 1.15. Kauffman's example after a detour.

Exercise 1.17. *Show that the diagram in Fig. 1.14 and by extension the diagram in Fig. 1.15 represent the trivial knot. Hint: The diagram lies on a sphere.*

1.7.2 *The fourth R move*

The other type of interactive move is R_4. In principle, the move, denoted $R_4(a, b, c, d)$, involves four tags $a, b, c,$ and d and is illustrated as the top move in Fig. 1.16. In practise, however, as we will see later when we come to consider virtualisation, singular knots and free knots, the move only involves two tags and is a sort of commutativity/anticommutativity. This is illustrated at the bottom of Fig. 1.16. The tag a swaps with the tag b and possibly becomes inverted.

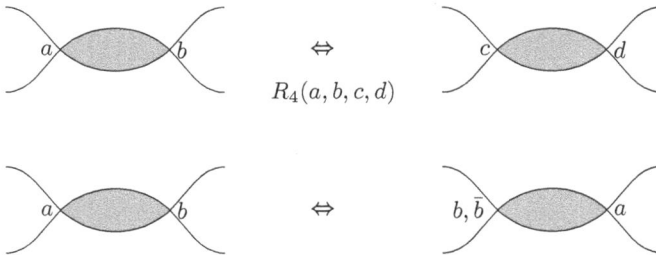

Fig. 1.16. R_4 moves.

1.8 Properties of various tags

Figure 1.17 indicates, by a tick, which R moves are allowed for various named tags acting on their own. Clearly, R_4 moves are redundant. We know or should know about the classical or real knots. As for the others, we will consider their associated theories in the following chapter.

Combinatorial Knot Theory

tag name	glyph	tag	R_1	R_2	R_3	
shadow		o				
flat		f	$\checkmark$	$\checkmark$	$\checkmark$	
ribbon		f^ϕ		$\checkmark$	$\checkmark$	
doodle		d	$\checkmark$	$\checkmark$		2d
virtual		v	$\checkmark$	$\checkmark$	$\checkmark$	
weld		w	$\checkmark$	$\checkmark$	$\checkmark$	
free		F	$\checkmark$	$\checkmark$	$\checkmark$	
homotopy		H	$\checkmark$	$\checkmark$	$\checkmark$	
+real		r	$\checkmark$	$\checkmark$	$\checkmark$	
-real		$\bar{r}$	$\checkmark$	$\checkmark$	$\checkmark$	3d
+singular		s		$\checkmark$		
−singular		$\bar{s}$		$\checkmark$		

Fig. 1.17. Allowed R moves.

Chapter 2

Generalised Knot Theories

Knit your hearts with an unsliping knot.
— Anthony and Cleopatra

2.1 Introduction

In the last chapter, we defined a **general knot theory** by diagrams and a number of allowable R moves acting on these diagrams. So, a generalised knot theory can be considered as an infinite graph with diagrams as vertices and R moves as edges. A knot in this theory is just a connected component of the graph. We now look at various examples of knot theories and their division into types.

The first binary division of theories is into simple and compound theories. In a **simple** theory, there is only one tag (and its inverse, if distinct). Classical knots are simple, as the diagrams only have the one tag r and its inverse $\bar{r}$. Theories with more than one type of tag are **compound** theories.

Another binary division is into 2-dimensional and 3-dimensional. A **2-dimensional theory** has $a = \bar{a}$ for all tags a.

A **3-dimensional** theory has $a \neq \bar{a}$ for at least one tag.

Finally, we split theories into **planar** and **non-planar** theories. Planar theories are defined by curves on the sphere. Non-planar theories are defined on any closed oriented surface.

2.2 Knots on a surface

We write $(a, b, \ldots)$ for the planar knot theory which is defined by planar diagrams tagged by $a, \bar{a}, b, \bar{b}, \ldots$ For non-planar theories, we use the notation, $((a, b, \ldots))$, but with the following proviso. We can think of the closed orientable surface of genus n as the boundary of a ball to which n handles have been attached. Around the centre of each handle is a **belt** meridian indicated by the letter b in Fig. 2.1.

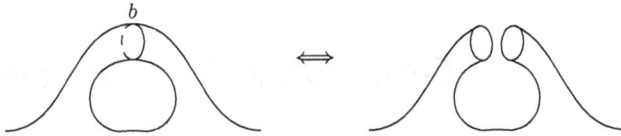

Fig. 2.1. Null surgery.

If no part of the diagram crosses b, then we can cut along b and glue in two disks as indicated in the right image of Fig. 2.1. Alternatively, we can do the reverse. These are called **null surgeries**, null because they do not impinge on the diagram.

We may sum up as follows. Two diagrams in $((a, b, \ldots))$ represent the same knot if they are related by a sequence of the following:

(1) homeomorphisms of the ambient surfaces which preserve the orientations and diagrams,
(2) R moves,
(3) null surgeries.

We shall see in a later chapter that knots on surfaces of genus greater than zero are intimately involved with planar virtual knots.

In what follows, a number of knots are presented as non-trivial or different from one another. These facts may need some algebraic invariants which will be the subject of the following volume. However, at the end of the book, we will introduce some simple invariants which will do the job.

2.3 Simple knot theories

In this section, we consider the theories generated by diagrams on surfaces with one tag.

The entry point for everyone who studies knots is the theory of classical knots, (r), which can be thought of as closed curves in space. They are simple (because there is only one tag (r) and 3-dimensional (because $\bar{r} \neq r$). The R moves allowed are the usual $R_1(r), R_2(r)$ and $R_3(r, r, r)$. This theory has a wide variety of literature and so will not be discussed further in this book except when it impinges on other theories. The theory $((r))$ can be interpreted as knots in $S \times I$ where S is a surface and I is the unit interval. This is the theory of **virtual knots** and will be considered in the following chapter.

2.4 2-dimensional simple theories

Classical knot theory is the only simple theory which is 3-dimensional. We now look at three different simple 2-dimensional theories. So, if a is the sole tag, then $a = \bar{a}$. For 2-dimensional knot diagrams, if the context is clear, it will not be necessary to indicate the tags in figures.

2.5 Flat knots, (f), $((f))$

The crossing points of these diagrams are labelled by the **flat tag**, f, which satisfies $R_1(f), R_2(f)$ and $R_3(f, f, f)$ The knots are the free homotopy type of the underlying curve on the surface. The homotopies which achieve this can be split up by an isotopy into a sequence of R moves. For planar diagrams, the knots are just equivalent to disjoint circles in the plane.

On a torus, the **flat Hopf**, consisting of a meridian and a longitude, represents a non-trivial flat knot with two components. Interestingly, there are no non-trivial examples on a torus with one component. The argument goes as follows: a curve on a torus with one component is homotopic to an integer multiple of a simple closed curve, γ. Let N be an annular neighbourhood of γ. So, we can assume that our one-component flat knot is contained within the interior of N. Now, surgery on one of the boundary components of N reduces the torus to the sphere.

In order to get a non-trivial one-component flat knot, we need to go to a surface of genus two. Here lies the flat Kishino knot.

We shall describe it and its lifted mother knot later. Moreover, we will develop a few simple tools to prove that it is non-trivial.

2.6   Doodles, (d), $((d))^5$

Planar flat knots are not very interesting, but we can make them interesting by restricting the number of R moves allowed. The first example excludes the R_3 move which gives us doodles, in particular planar doodles defined by curves in the plane, (d). Doodles on surfaces of higher genus will be considered later.

The **doodle tag**, d, satisfies $R_1(d)$ and $R_2(d)$. So, the triple point creating R_3 is not allowed. We shall see later that any doodle has a unique minimal representative diagram with no monogons or bigons. The Borromean rings illustrated earlier is the smallest non-trivial planar doodle. Its minimal representative has six crossing points. The poppy illustrated in Fig. 2.2 has one component and eight crossing points.

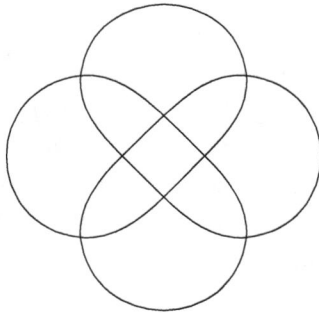

Fig. 2.2.   The poppy.

Exercise 2.1. *Show that there is no planar doodle having no monogons or bigons with a minimum of seven crossings.*

We will return to doodles at a later stage.

---

[5]The concept of a doodle is due to Fenn and Taylor [70]. In their definition, component curves were simple and so lacked self-crossings. This more general definition is due to Khovanov [156].

2.7 Regular flat knots, (h), $((h))$

In this section, all curves will lie in the plane. Suppose we consider curves in the plane under R moves but now exclude the R_1 move. Let h be a tag which satisfies $R_2(h)$ and $R_3(h, h, h)$. Any planar curve which is acted upon be a sequence of these moves is called a **regular homotopy**. During a regular homotopy, the tangent at each point of the curve varies continuosly and is never zero. Alternatively, we can think of a regular flat knot diagram as a ribbon pressed into the plane with no kinks and with a finite number of double squares where one part of the ribbon being a square presses down on another.

The resulting knots are called **regular flat knots**. Since different components of a planar curve can be separated by a regular homotopy, we need only consider regular flat knots with one component. It turns out that these are classified by an integer.

Let the curve $\alpha : S^1 \to \mathbb{R}^2$, representing a regular flat knot, have a continuously varying non-zero tangent vector at every point. By adjusting the parametrisation, we can assume that each tangent vector has unit length. So, the tangent, α' defines a map from S^1 to itself. The homotopy type of this map is an integer called the **turning number**[6] of the curve. For example, the turning number of the standard circle is ± 1 depending on whether it is oriented anticlockwise or clockwise and the turning number of the number eight is zero.

Exercise 2.2. *To familiarise yourself with this concept, draw a curve of turning number 2 and one of turning number -2.*

The turning number is clearly an invariant of regular homotopy and, as we will now see, also a complete invariant. The following theorem involves calculus. If this gives you an attack of the vapours, you can jump forward to a different proof.

Theorem 2.1 (Whitney–Graustein[7]**).** *Regular homotopy classes of maps of a circle into the plane are classified by their turning numbers.*

---

[6]Whitney calls this the rotation number.
[7]In his paper, Whitney attributes the idea to Graustein.

Proof. We have already seen that turning numbers are invariant. The key idea in Whitney's proof of the converse is the following observation. Let $f : [0, 1] \to S^1$ be a continuous path in the circle and let p be a point in the plane. The integral

$$\alpha(t) = p + \int_0^t f(u) du$$

$t \in [0, 1]$ defines a smooth path in the plane with a unit tangent at every point and with initial point the constant of integration, p. This will be a closed curve based at $p = \alpha(0)$ with a continuously varying tangent if

$$f(1) = f(0) \quad \text{and} \quad \int_0^1 f(u) du = 0$$

The second equation can be summed up as **the average value of f is zero**.

Let $\alpha_0, \alpha_1 : [0, 1] \to \mathbb{C}$ be smooth curves in the plane with unit tangents at every point and having the same turning number, N. Let f_0 and f_1 be the unit tangents of α_0 and α_1 defined by differentiation. We may assume by a translation and a rotation that $\alpha_0(0) = \alpha_1(0) = p$ and $f_0(0) = f_1(0)$. Then f_0 and f_1 are homotopic representatives of $N \in \pi_1 S^1 \cong \mathbb{Z}$. So, there is a homotopy $f_s, 0 \leq s \leq 1$ between them. We would like to take this function and apply the observation above. However, we cannot be sure that the second equation is satisfied. But if we replace the homotopy by the translation

$$f_s - \int_0^1 f_s(u) du$$

then the second equation is satisfied. Full details can be found in the work by Whitney [267]. □

There is a simpler more conceptual proof which doesn't use analysis, as we shall show shortly. However, the interesting and important fact about the above proof is that it is capable of being generalised and is the forerunner of the results by Smale [244] and others on the immersions and regular homotopies of spheres.

The **standard** immersed curve, C_N, of turning number N is defined as follows. Take a circle oriented anticlockwise and attach

$N - 1$ curls (monogons) to the inside if $N > 0$. If $N \leq 0$, attach $|N| + 1$ curls to the outside. In Fig. 2.3, C_3 and C_{-2} are illustrated.

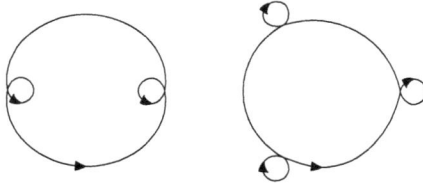

Fig. 2.3. The standard curves, C_3 and C_{-2}.

The following result reproves Theorem 3.2.

Theorem 2.2. *Any curve with turning number N is regularly homotopic to the standard curve C_N.*

Proof. The curve can be thickened to an immersed ribbon annulus by considering the trace of small normals to the left and right of the tangent at any point. The curve is now the centre line of the annulus and the boundary of the annulus are the two curves traced out by the endpoints of the two normals. Figure 2.4 illustrates this procedure for the trefoil.

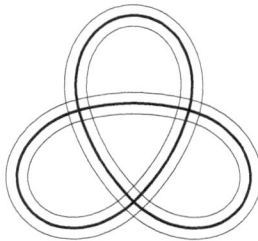

Fig. 2.4. A thickened trefoil curve.

We can homotop the circle to a standard embedded anticlockwise oriented circle by imagining that this is a flat curve and using a sequence of the three f moves. This would be a regular homotopy but for the $R_1(f)$ moves. However, we can get round this by allowing the ribbon to do the $R_1(f)$ moves while the curve performs a regular homotopy within the ribbon. This is possible as two opposite monogons can be cancelled by a sequence of $R_3(f)$ moves and $R_2(f)$ moves (Fig. 2.5).

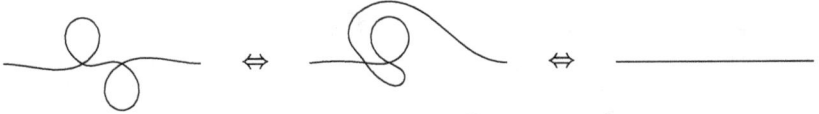

Fig. 2.5. Cancelling opposite monogons.

If the R_1 move is an expansion, then the curve does the opposite expansion and the new curve is etched in the ribbon, as illustrated in Fig. 2.6.

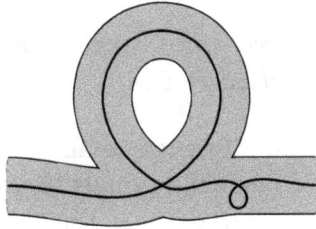

Fig. 2.6. The result of an R_1 expansion.

If the R_1 move is a collapse, then the ribbon also collapses, but the curve does not and just records the monogon in the ribbon (Fig. 2.7).

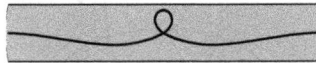

Fig. 2.7. The result of an R_1 collapse.

So, the curve is regularly homotopic to a circle with a number of monogons attached, some on the inside and some on the outside.

We can assume that all the monogons are all on the outside or all on the inside of the circle and so wind in the same direction because two adjacent monogons on different sides can be cancelled by the trick illustrated in Fig. 2.5, which removes the two self intersections. The result is C_N.

So, any curve with turning number N is regularly homotopic to C_N, a circle with monogons attached and all oriented the same way.

Hence, any curve with turning number N is regularly homotopic to any other curve with turning number N. □

2.8 Calculating the turning number

The second proof of 2.7.2 suggests a way of calculating the turning number from the diagram. Suppose we consider stationary points (max/min) of the y-coordinate restricted to the curve. A left to right minimum is called a **smile** and a left to right maximum is called a **frown**. Right to left stationary points are ignored (Fig. 2.8). For example, $C3$ has three smiles and no frowns, whereas C_{-2} has one smile and three frowns.

Fig. 2.8. Right directed minimum (smile) and maximum (frown).

Let N_+ be the number of smiles and let N_- be the number of frowns.

Theorem 2.3. *The turning number of a curve is* $N = N_+ - N_-$.

Proof. We can adjust the R moves of a regular homotopy so that N_+ and N_- are unchanged. The only change comes now with a general homeomorphism. This might introduce neighbouring smiles and frowns. But the difference is unaltered as shown in Fig. 2.9.

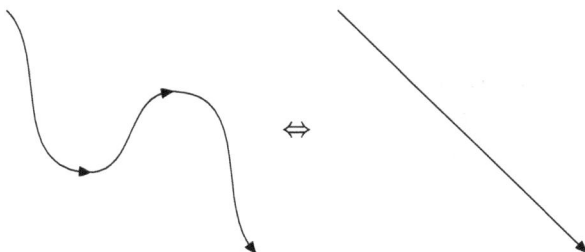

$\Leftrightarrow$

Fig. 2.9. A smile cancels a frown.

Now, check the answer with the standard curve C_N. □

Exercise 2.3 (Whitney [267]). *Let α be a smooth curve. A* **starting point**, *z_0, on the curve is one where the curve lies on one side of the tangent line at z_0. If it lies on the left side of the tangent line, put $\mu = 1$. If it lies on the right side, put $\mu = -1$. Orient the crossings of the curve like little coordinate systems. So, at a crossing point, one tangent vector points in the x-direction (East) and the other points in the y-direction (North). Start from z_0 and continue round the curve in a positive direction. A crossing is deemed* **positive** *if the first encounter is from the South and is* **negative** *if from the West. Note that this definition depends on the starting point z_0. Let M_+ be the number of positive crossings and let M_- be the number of negative crossings. Show that the turning number is given by $N = \mu + M_+ - M_-$.*

2.9 Winding number

A similar concept to the turning number is the **winding number**. Let p be a point of the plane not on a curve. Let q be a point on one component α. As q traverses α, the unit vector $(q - p)/|q - p|$ defines a map, called the **winding function**, from the circle to itself. The algebraic number of times the winding function wraps round is called the **winding number** at p. Write $w = w(p; \alpha)$ for this integer. This is the element of the fundamental group defined by the winding function.

If p is moved around the interior of a region, then w is unchanged. If p crosses an edge into a new region, then w changes by ± 1. To see this and to determine the sign, imagine that p_1 and p_2 are on opposite sides of an edge which we can think of as the y-axis oriented up the page with p_1 on the left and p_2 on the right. Push a little finger out from the edge to enclose p_2. This is topologically the same as moving p_2 to p_1. The effect on S is to compose it with a small anticlockwise circle around p_2. So, w becomes $w + 1$ on crossing an arc from left to right. Moving p in the opposite direction removes 1 from w.

Exercise 2.4. *Consider the winding number of points in the regions of a single curve in the plane. If the crossing points of the curve are oriented in the normal way as small coordinate system, then the* *neighbouring regions can be named North West, North East, South East and South West. Show that a point in one of these regions has winding number $n, n+1, n+2, n+1$, respectively, where n is some integer.*

If D is some planar diagram, then the **total winding number** is the sum of the individual winding numbers for each component.

Later, we will consider braided diagrams which have one region A with maximum total winding number and one region B with zero total winding number such that the intervening regions have monotonically decreasing total winding number.

Exercise 2.5. *Colour regions of a planar diagram which have odd total winding number black. Colour the rest white. Show that this defines a chess board colouring of the diagram.*

The following table sums up the simple 2-dimensional theories covered so far. Since they are 2-dimensional, all tags satisfy, $\bar{x} = x$. Moreover, they all have the same glyph that is a simple crossing in the plane.

Name	Tag	R_1	R_2	R_3
Shadow	o			
Flat	f	✓	✓	✓
Doodle	d	✓	✓	
Regular	h		✓	✓

2.10 Compound knots

In principle, compound knots can have any number of tags, but in this book, we will restrict to just two. A detailed discussion of compound knot theories with the virtual tag will be the subject of Chapter 3.

Recall the following important concept in the theory of compound knots. We say that a tag x **dominates** a tag y if the R move, $R_3(x, x, y)$, is satisfied. We can think of this as the crossing tagged y passing past an arc with two x tags, see Fig. 2.10. If x dominates y, we write $x \succ y$.

Fig. 2.10.　x dominates y.

For example, r or $\bar{r}$ dominates r and $\bar{r}$.

Recall that if x dominates y, then x also dominates $\bar{y}$.

2.11　Singular knot theory, (r, s)

A singular knot diagram has classical crossings and the two singular crossings s and $\bar{s}$ illustrated in Fig. 2.11.

Fig. 2.11.　Tags s and $\bar{s}$.

The glyph for s is a black disk and the glyph for $\bar{s}$ is a white disk (Fig. 2.11). It is important to realise that these are disks with positive radii not just points. The arcs which meet the disk are embedded as orthogonal diagonals. Their motion in space is constrained by this fact. So, if a disk rotates so that its front is now its back, then the disk will twist the neighbouring edges and this is covered by the fourth R move considered later.

To the classical Reidemeister, r move, $R_2(s)$ or $R_2(\bar{s})$, is added (Fig. 2.12).

Fig. 2.12.　Black and white singular disks cancel.

Both r and $\bar{r}$ dominate s and $\bar{s}$. So, $R_3(r, r, s^{\pm 1})$ and $R_3(\bar{r}, \bar{r}, s^{\pm 1})$ are satisfied. Only black disks are illustrated in Figs. 2.13–2.15 and they could be replaced by white disks.

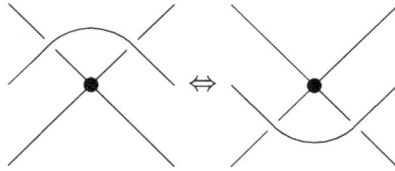

Fig. 2.13. $r \succ s$. Moving an arc over a singular disk.

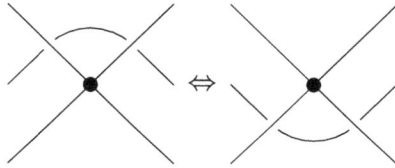

Fig. 2.14. $\bar{r} \succ s$. Moving an arc under a singular disk.

The fourth condition mentioned earlier is an $R_4(r, s^{\pm}, s^{\pm}, r)$ move illustrated in Fig. 2.15.

Fig. 2.15. Twisting a disk.

The orientations of the illustrated arcs are arbitrary.

The geometric interpretation of a singular knot diagram is the following. Imagine several smooth closed curves in space which are embedded except at a finite number of places where they cross a black or white disk transversely. This physical object can now move about freely in space, giving rise to all the moves indicated in figures above except $R_2(s)$ and $R_2(\bar{s})$, which do not seem to have a physical interpretation.

2.12 Homotopy knots $(r, \pm r)$

Homotopy knot theory is defined by planar diagrams with classic crossings and a shape shifter called a $\pm$ **crossing**. This new kind of crossing is both a positive real crossing and a negative real crossing.

In other words, the over and under points of a real crossing can arbitrarily interchange so that a positive crossing becomes negative and vice versa. Its tag is $\pm r$ and its glyph is a hole where the crossing is. A diagram has the following rules: crossings from the same component are tagged with $\pm r$ and crossings from two different crossings have standard r or $\bar{r}$ tags (Fig. 2.16).

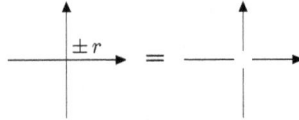

Fig. 2.16. The $\pm$ crossing of a single component.

We can visualise a homotopy knot diagram in $\mathbb{R}^3$ as a curve in space by lifting the diagram into the third dimension. The real crossings have their usual interpretation. The $\pm$ crossings can be made into r or $\bar{r}$ crossings at whim. The two double points then project onto the $\pm$ crossings. We can then move the curve in space with one crucial difference. Arcs from the same component can pass through each other. Arcs from different components cannot.

Lemma 2.1. *We can assume that the components of a homotopy knot realised in $\mathbb{R}^3$ are unknotted.*

Proof. Let the diagram plane be a horizontal (x, y)-plane and let z be the height function. Pick a starting point X, which is not a crossing, on a component c. The idea is to change the knot in space so that as we move along c, the height function steadily decreases. On encountering a pre-image of a self double point, if it is the first time, then we continue along all the time making sure that z is decreasing. If it is the second time and an undercrossing point, then there is nothing to do. If it is the second time and an overcrossing point, then push the overcrossing point past the undercrossing point so that their roles are reversed. Eventually, we return to X. Chose a point X_0 under X and join them by a vertical line which is now part of the knot. So, the component c decreases in height from X to X_0 which is then joined by an upward vertical segment to X. We must show that c is unknotted.

Consider the projection onto the (x, z)-plane. After a small adjustment, the image is a segment and a monotonically decreasing arc which joins the ends of the segment and which crosses the segment at a finite number of places. This is clearly the image of an unknotted curve. □

Illustrated in Fig. 2.17 are two examples.

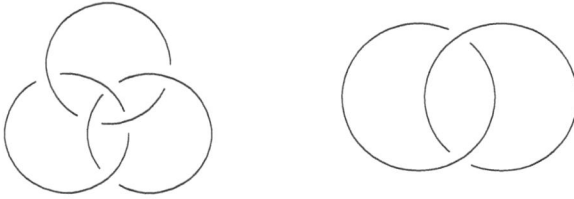

Fig. 2.17.   Non-trivial homotopy knots: The Borromean and Hopf links.

Exercise 2.6. *Show that any height function on a component of a knot has an equal number of maxima and minima.*

Exercise 2.7. *Show that the trefoil knot has a height function with two maxima and two minima.*

Exercise 2.8. *The* **linking number***, μ_{12} of two components c_1 and c_2, of a knot is defined by the number of positive crossings of c_1 and c_2 minus the number of negative crossings all divided by 2. Show the following:*

(1) *μ_{12} is invariant under the R moves.*
(2) *μ_{12} is an invariant of homotopy knots.*
(3) *Find an example of a two-component knot with arbitrary μ_{12}.*

The linking number μ_{12} is an example of a **Milnor number**. For the Hopf link, $\mu_{12} = \pm 1$. For the Borromean rings, μ_{12}, μ_{23} and μ_{31} are all zero, but another Milnor number $\mu_{123} = \pm 1$. We will consider the Milnor numbers in the following volume.

The contents of this chapter are summed up in Fig. 2.18. The tag x in the dominance column stands for any tag. A symbol such as $x \leftrightarrow y$ means that the R_4 move which interchanges x, y is applied.

Tag name	Glyph	Symbol	R_1	R_2	R_3	R_4	Dominance
shadow		o					
flat		f	✓	✓	✓		
doodle		d	✓	✓			
regular		h		✓	✓		
virtual		v	✓	✓	✓		$v \succ x$
weld		w	✓	✓	✓		$w \succ x$ $r \succ w$
free		F	✓	✓	✓	$F \leftrightarrow v$	$w \succ x$ $r \succ w$
homotopy		H	✓	✓	✓		$H \succ r, \bar{r}$ $r, \bar{r} \succ H$
+real		r	✓	✓			$r \succ r, \bar{r}$
−real		$\bar{r}$	✓	✓			$\bar{r} \succ r, \bar{r}$
+singular		s					$s \leftrightarrow r, \bar{r}\, r, \bar{r} \succ s$
−singular		$\bar{s}$					$\bar{s} \leftrightarrow r, \bar{r}\, r, \bar{r} \succ \bar{s}$

(rows shadow–homotopy braced as "2d"; rows +real–−singular braced as "3d")

Fig. 2.18.   Tag table.

Chapter 3

Chord Diagrams

3.1   Chord diagrams

A **chord diagram** is a collection of disjoint circles and **chords**. The chords can either be **internal** with end points on one circle or **external** with end points on two circles. The only intersection of the chords with the circles are the chord endpoints. Different chords may appear to cross but are disjoint.

Let $\alpha : \bigsqcup S_i^1 \to S^2$ be a curve defining the shadow of a diagram, $|D|$, in some knot theory. Suppose $x, y \in \bigsqcup S_i^1$ are distinct points which define a crossing point, c, so $\alpha(x) = \alpha(y) = c$. Then provided c is not a virtual crossing, join x to y by a path (**chord**). The result is a chord diagram.

We can add bells and whistles to such a chord diagram by labelling the chord by the same tag as c. If the crossing is a real crossing tagged by r or $\bar{r}$, then we orient the chord to make an **arrow chord** (see Fig. 3.1). By convention, the orientation is from the overcrossing point to the undercrossing point of the crossing. The **sign** of this chord is $+$ or $-$ and corresponds to r and $\bar{r}$, respectively.

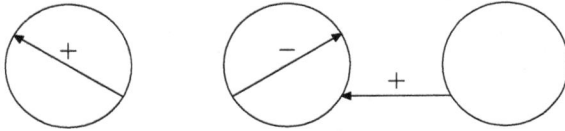

Fig. 3.1.   Arrow chords.

The chord diagram of the left-handed trefoil is illustrated in Fig. 3.2. The three arrow chords only *appear* to cross in the centre of the circle. All three are disjoint, are labelled with $-$, and so correspond to $\bar{r}$.

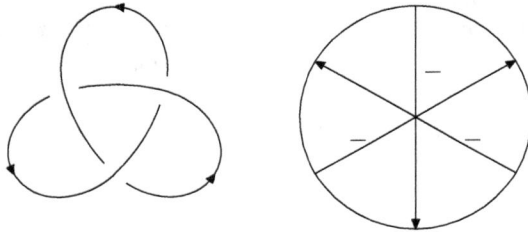

Fig. 3.2.   Left-handed trefoil and chord diagram.

If we mix classical crossings with others, we get something like the illustration in Fig. 3.3.

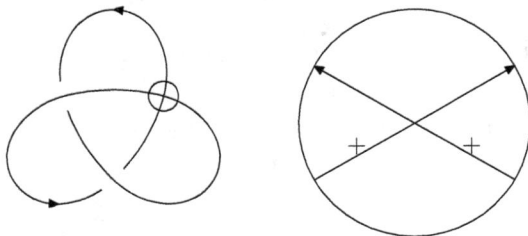

Fig. 3.3.   Right-handed virtual trefoil and chord diagram.

Note that there is **no chord** corresponding to the virtual crossing.

3.1.1 *The actions of R moves on chord diagrams*

We now look at how the R moves alter the corresponding chord diagrams.

In the illustrative figures, the dotted part of the circle is fixed while the solid part or parts change between left and right of the figure. The chords may be oriented in case we are dealing with real crossings.

3.1.2 *The result of an R_1 move on a chord diagram*

The R_1 move (see Fig. 3.4) introduces or deletes a chord which divides the circle into an arc with no chord ends and the rest.

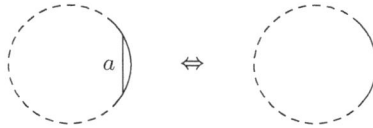

Fig. 3.4. Action of R_1 on a chord diagram.

3.1.3 *The result of an R_2 move on a chord diagram*

For the next two actions of R_2 and R_3, we assume that the knot has only one component. For more than one component, the reader can easily supply their own figures and arguments.

Figure 3.5 shows the action of a non-parallel R_2 move from the middle to the left circle and a parallel R_2 move from the middle to the right circle.

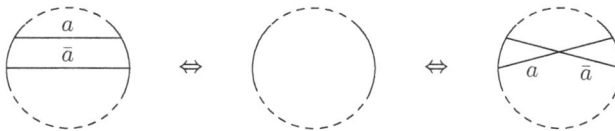

Fig. 3.5. Action of R_2 on a chord diagram.

3.1.4 *The result of an R_3 move on a chord diagram*

Figure 3.6 shows the action of $R_3(r, r, \bar{r})$. This can be interpreted as the domination of the classical crossing r over $\bar{r}$. Similar figures could indicate other classical dominance. The action leaves the orientation and the sign of the arrow chords unchanged but moves one chord which intersects the two other chords so that they do not intersect and simultaneously crosses the previous uncrossed chords.

Fig. 3.6. Action of $R_3(r, r, \bar{r})$ on an arrow chord diagram.

The diagram assumes that, following the orientation of the circle, knot arcs connect from the top to the middle to the bottom. The reader is invited to describe a chord diagram change when the arcs move from top to bottom to middle. The effect is similar but moves from right to left rather than left to right.

Figure 3.7 shows the action of $R_3(a, b, c)$. Remember that if one of the tags is r or $\bar{r}$, then the chord needs to be changed into an arrow chord corresponding to a classical crossing.

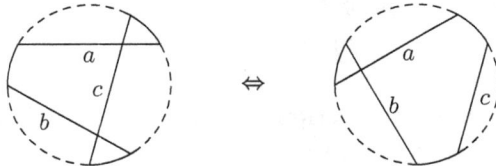

Fig. 3.7. Action of $R_3(a, b, c)$ on a chord diagram.

3.1.5 *The result of an R_4 move on a chord diagram*

There are two types of R_4 actions to consider. In one type, none of the two tags involved is a virtual tag, v. Then the effect of the action of R_4 is to swap the tags on the two chords involved possibly inverting one tag. An example of this is when the tags are the singular tag, s, and the real tag, r. The tags are swapped but not inverted.

The other type has one tag equal to v, the virtual crossing. But the virtual crossing is not there, so it has no associated chord.

It follows that there is only one chord labelled by a tag, a, say. If a is not the real tag r, then after the R_4 move, the chord will be unchanged except possibly the tag will change to $\bar{a}$. If a is the real tag, then the chord is directed and the move reverses this direction and possibly changes r to $\bar{r}$.

It follows that if we have a knot theory, $\mathcal{K}$, and a knot diagram, D, tagged by the theory, then D determines a chord diagram, D', say. The R moves on D can be mirrored by R moves on D', giving a bijective correspondence between knot diagrams and chord diagrams which come from knot diagrams. Moreover, this bijection is invariant under the R moves.

3.1.6 *Unrealisable chord diagrams*

Corresponding to every knot diagram is a chord diagram. This chord diagram is called **realisable** in the appropriate knot theory. But not every chord diagram is realisable in the given theory. In Fig. 3.8, two unrealisable chord diagrams are shown. The diagram on the left is unrealisable as a classic knot. The diagram on the right is unrealisable as a flat knot with two components.

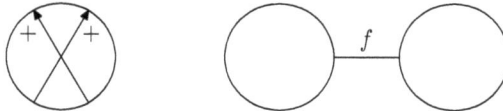

Fig. 3.8. Unrealisable chord diagrams.

3.1.7 *Invariants of chord diagrams*

To understand why some chord diagrams such as those in Fig. 3.8 are unrealisable in their theory, we will introduce some simple invariants. Later in the book, we will look at other simple invariants.

Let χ be a chord joining two points on a circle, C. The end points of χ divide C into two open arcs, C' and C''. We say that χ is an **odd** chord if C' (or C'') has an odd number of chord ends from other internal chords of C. Otherwise, χ is said to be **even**. For example, the left-hand chord diagram in Fig. 3.8 has two odd chords.

Let D be a chord diagram with arrow chords. The **writhe**, $\mathcal{W}(D)$, of D is defined to be the algebraic sum of all the signs of the arrow chords. So, for example, the left-hand trefoil in Fig. 3.2 has writhe -3 and the right-hand virtual trefoil in Fig. 3.3 has writhe 2. The writhe is invariant under R_2, R_3, R_4 and changes by ± 1 under R_1.

Define the (real) **odd writhe**, $\mathcal{W}_o(D)$, to be the algebraic sum of the signs of the odd arrow chords. So, the example on the left of Fig. 3.8 has odd writhe 2.

Now, let D be a chord diagram with chords tagged by f, the flat tag. Define the (flat) **odd writhe**, $\mathcal{W}_{of}$, to be the number of odd flat chords which are tagged by the flat tag, f, taken mod 2. Both odd writhes are invariant under all the R moves.

Let C_1 and C_2 be two circles in a chord diagram, D. The (real) **linking number** $\mu_{12}(D)$ is the algebraic sum of the sign of the external arrow chords joining C_1 and C_2 divided by 2. The **flat linking number**, $\phi_{12}(D)$, is the sum, mod 2, of the external chords joining C_1 and C_2 tagged by the flat tag, f. So, the flat linking number of the example to the right of Fig. 3.7 is 1.

Theorem 3.1. *Let K be a real knot diagram and let D be the associated chord diagram. Then*

(1) $\mathcal{W}_o(D) = 0$;

(2) μ_{12} *is an integer for any pair of circles C_1 and C_2 of D.*

Proof. Let c be a crossing of K involving one component. Then it is associated with an internal chord, χ, dividing a circle C into two open arcs, C', C''. The image of C' in the plane is a loop which must cross an even number of arcs, so χ is even and there are no odd chords.

Now, suppose that K_1 and K_2 are the components of K corresponding to the circles C_1 and C_2 of the chord diagram. Then K_1 and K_2 cross an even number of times, and so, the sum of the external chords joining them is divisible by two. $\qquad\square$

Theorem 3.2. *Let K be a flat knot diagram and let D be the associated chord diagram. Then*

(1) $\mathcal{W}_{of}(D) = 0$;
(2) ϕ_{12} *is zero for any pair of circles C_1 and C_2 of D.*

Proof. The proof is similar to 3.4.4. □

It follows that the example on the right of Fig. 3.7 cannot be the chord diagram of a flat knot.

Chapter 4

Virtual Crossings and Virtual Knots

This sort of crossing is called virtual. It comes in only one flavor. You cannot switch over and under in a virtual crossing. However, the idea is not that a virtual crossing is just an ordinary graphical vertex. Rather, the idea is that the virtual crossing is not really there.

— Louis Kauffman

4.1 Virtual knots and crossings

The numbers $0, -1, i = \sqrt{-1}$ were invented to solve a problem and these inventions were controversial at the time. The symbol i for imaginary reflects this. In a similar fashion, the name *virtual* suggests that such a crossing does not exist. Such vacuous crossings were invented to allow the path of a curve in the plane to cross another path without increasing the number of real crossings.

We met the virtual crossing and tag earlier in the book. It occupies a central position in combinatorial knot theory. In this section, we will bring out its importance as a link between planar knots and those on other surfaces. In particular, we will show how it comes to the rescue of unrealisable chord diagrams and realises them as a planar diagram with added virtual crossings. Its invention leads to other concepts, such as virtualisation, forbidden moves, free knots and welds. These will be treated in this chapter. Finally, we will show how virtual knots are the correct vehicle for knots on surfaces of higher genus.

The tag for a virtual crossing is v, but it also has a built-in glyph in the same way that the real crossing r has. The symbol is a small circle surrounding the crossing. This is illustrated in Fig. 4.1.

Fig. 4.1.   A virtual crossing and its glyph.

The R moves obeyed by v are the same as those of the flat tag f, namely $v = \bar{v}, R_1(v), R_2(v)$ and $R_3(v, v, v)$. The difference is the following:

Virtual tags dominate everything, even real crossings. So, $R_3(v, x, v)$ is true for all x as illustrated in Fig. 4.2.

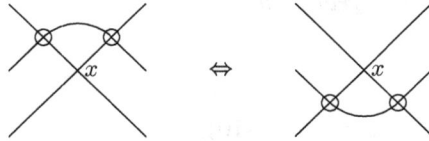

Fig. 4.2.   Virtual crossings dominate all.

Call an arc of a diagram which only passes through virtual crossings a **virtual arc**. Keeping the ends of a virtual arc fixed, we can allow the interior to wander where it will, provided possible new virtual crossings are introduced or old ones deleted using the R moves and dominance. This move is an example of the detour move considered earlier.

4.1.1 *Adding a virtual tag to a theory*

To realize chord diagrams, we need the following definition. Let $\mathcal{K}$ be a knot theory represented by planar diagrams tagged by a collection of tags, $\mathcal{A}$, which exclude the virtual tag. So, $\mathcal{K} = (\mathcal{A})$ using the earlier notation. Consider the theory, which we write as

$\mathcal{K} + v$ or $(\mathcal{A}, v)$, consisting of diagrams tagged by $\mathcal{A}$ and the virtual tag, v, which dominates everything. We call this theory the **virtual $\mathcal{K}$ theory**.

Virtual real knots, (r, v), being the first to be considered, are usually just called **virtual knots**. Other virtual knot theories in the literature are **virtual flat knots**, (f, v), and **virtual doodles**, (d, v), but in principle, any planar non-virtual knot theory can be appended with the adjective, *virtual*.

4.1.2 *Realising unrealisable chord diagrams*

If virtual crossings are allowed, then the unrealisable chord diagrams of Fig. 3.8 can be realised as a virtual knot, the **virtual right-hand trefoil** and a virtual flat knot, the **virtual flat Hopf link**, see Fig. 4.3.

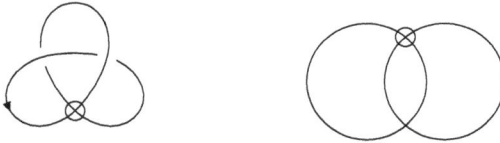

Fig. 4.3. Realizing the unrealisable chord diagrams.

Theorem 4.1. *Any chord diagram in a knot theory $\mathcal{K}$ can be realised as a planar diagram in the virtual theory $\mathcal{K} + v$. Moreover, this realisation is unique up to detour moves.*

Proof. Although we have announced this as a theorem, the proof is very easy and is a natural progression of the definitions.

Realizing chord diagrams is achieved by placing the crossings with their tags and orientation disjointly on the sphere and joining up the ends by arbitrary arcs. Where these arcs cross, we have virtual crossings. By detour moves of virtual arcs, it is clear that the results always represent the same virtual knot. □

4.1.3 *Virtualisation*

This R_4 move can be utilised to change a real crossing and post two virtual crossings on either side. This move is called **(positive) virtualisation** by Kauffman.

Fig. 4.4.   Virtualisation.

The effect of virtualisation on the chord diagram is to reverse the direction of the chord, but the sign is unchanged. To see what R_4 move is allowed, introduce two virtual crossings on one side of the real crossing by an R_2 move. Then apply an R_4 move involving the real crossing and the adjacent virtual crossing. This is virtualisation and is illustrated in Fig. 4.4. The red arc which previously ducked under the blue arc now rides over it.

If we apply virtualization to the chord diagram of the trefoil, then one of the arrow chords reverses to get what Kauffman calls the **virtualised trefoil**. It has no odd chords in its chord diagram. So, its odd writhe is zero, but yet cannot be realised as a real knot.

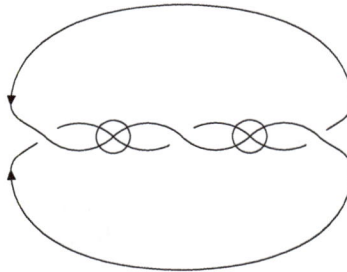

Fig. 4.5.   Virtualised trefoil.

We can also do a **(negative) virtualisation**. This involves reversing an arrow chord and also changing its sign. So, in Fig. 4.4, the blue arc would continue to cross the red arc above. If we apply this to the chord diagram of the trefoil, we get the **negative virtualised trefoil**.

Exercise 4.1. *Draw a negative virtualised trefoil.*

Exercise 4.2. *Show, up to orientation, that there are five different knots with three arrow chords in their chord diagram: the* **unknot** *with no crossing real or virtual, the* **classic trefoil** *with three real crossings, all of the same sign and no virtual crossings, the* **virtual**

trefoil *with two real crossing of the same sign and one virtual cross-ing, the* (**positive**) **virtualised trefoil** *with three real crossings of the same sign and two virtual crossings and the* (**negative**) **virtu-alised trefoil** *with three real crossings, two of the same sign and two virtual crossings.*

We saw earlier that given any non-trivial classical knot diagram, we can reverse some of the crossings to get the unknot. If instead we virtualise these crossings, then we would get the unknot except we have a number of virtual crossings in the way. It is an unsolved problem at the time of writing as to whether this new knot can ever be classical. If we knew the answer to this problem, then we might find a non-trivial one-component knot with trivial Jones polynomial. This, however, is consideration for a later book.[8]

It is a remarkable fact that virtual crossings, only considered so as to satisfy a planar problem, can be realised on a surface. We will consider this in Theorem 4.3.

In the meantime, consider the following oddities.

4.1.4 *Welded knots and the forbidden moves*

If we add to virtual knot theory the ability for a real tag to dom-inate a virtual crossing (now called a **weld**), we get the theory of welded knots. More precisely, the weld tag, w, satisfies $w = \bar{w}$, $R_1(w), R_2(w), R_3(w, w, w)$ and like the virtual tag, w dominates $r, \bar{r}$ and has the same glyph, but unlike the virtual tag, r dominates w. So, we have $R_3(w, w, r), R_3(w, w, \bar{r})$, and also $R_3(r, r, w)$ which we saw is equivalent to $R_3(r, w, \bar{r})$ and is illustrated in Fig. 4.6 if we assume all arcs are oriented from left to right. The last condition or motion is called the **first forbidden move**.

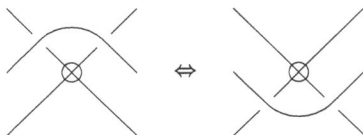

Fig. 4.6. The first forbidden move.

---

[8]There is now proof that this is impossible.

On the other hand, the **second forbidden move**, $R_3(\bar{r}.\bar{r}, w)$, is not allowed, i.e., an arc cannot move under a weld as in Fig. 4.7. Imagine that the weld is glued to the plane. We will show that if both forbidden moves are allowed, then the theory collapses.

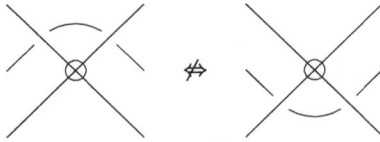

Fig. 4.7.   The second forbidden move.

Theorem 4.2 (Nelson). *Under the action of both forbidden moves, all virtual knots become trivial.*

Proof. If we allow both forbidden moves, then it is easy to see that the following transformations F_1 and F_2 illustrated in Figs. 4.8 and 4.9 are possible.

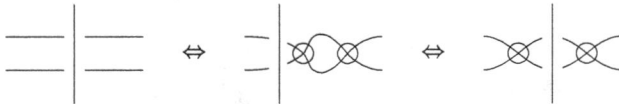

Fig. 4.8.   Transformation F_1.

To see this, create two virtual crossings by R_2 for virtual knots and then slide one of them under the vertical arc by the second forbidden move. A similar argument using the first forbidden move allows the transformation F_2 to be defined.

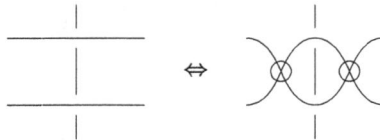

Fig. 4.9.   Transformation F_2.

We now use chord diagrams to finish the proof. In Fig. 4.10, the effect of F_1 and F_2 on the corresponding chord diagrams is to interchange two adjacent arrow heads and two adjacent arrow tails, respectively.

Fig. 4.10. The effect of F_1 and F_2 on the chord diagram.

However, this is not enough. We now need F_3, which interchanges an arrow head with an adjacent arrow tail. This process is illustrated in Fig. 4.11. It follows a sequence of Reidemeister 2, F_1, F_2, Reidemeister 3, and finally, Reidemeister 2 again.

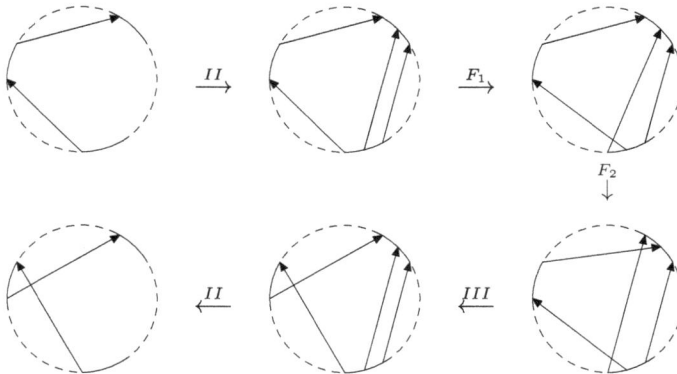

Fig. 4.11. Swapping heads and tails with F_3.

So, by reversing this sequence, any chord diagram can be reduced to one in which all the arrows are disjoint. This is clearly the unknot.

□

4.1.5 *Welded knots and tori*

The welded knots have a beautiful geometric interpretation as tori in 4-dimensional space.

We start by dividing $\mathbb{R}^2$ and $\mathbb{R}^4$ into stacked hyperspaces. Let the coordinates of $\mathbb{R}^2$ be (x, t) where the positive x coordinate is across the page from left to right and the positive t coordinate is up the page. For a fixed value of t, the line

$$L_t = \{(x,t)|x \in \mathbb{R}\}$$

is horizontal and as t varies over $\mathbb{R}$, these partition the plane. Let K be a planar diagram. We can slightly adjust K so that the sets $K \cap L_t$ consist of the following:

(i) a finite number of **regular points** where the diagram crosses L_t transversely;
(ii) a single maximum or minimum;
(iii) a single crossing.

Values of t where (i) occurs are called **regular**. Otherwise, they are **singular** values and the points are **singular** points.

Now, consider the partition of $\mathbb{R}^4$ into 3-dimensional **solids**,

$$M_t = \{(x,y,z,t)|(x,y,z) \in \mathbb{R}^3\}$$

We will associate a disk, $D_{(x,t)}$, in M_t to every regular point (x,t) in a continuous fashion so that at every singular point, the following occurs.

At a maximum/minimum, the singularity is a figure eight which either splits into two or is a single disk in the neighbouring regular values. See Fig. 4.12. The picture for a minimum is similar.

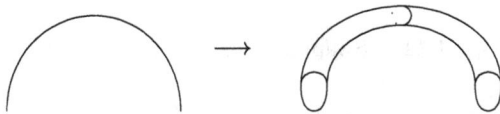

Fig. 4.12.   Interpretation of a maximum.

At a weld, the disks pass each other without meeting, see Fig. 4.13.

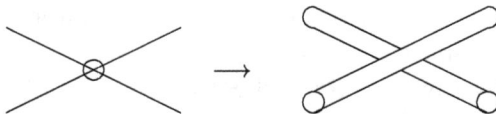

Fig. 4.13.   Interpretation of a weld.

At a real crossing, the disk at the singular point of the under arc lies in the interior of the disk at the singular point of the over arc, see Fig. 4.14.

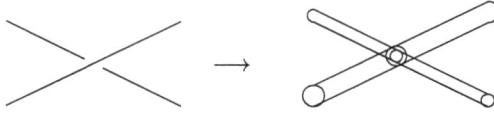

Fig. 4.14. Interpretation of a real crossing.

So, corresponding to a diagram of a welded knot, there is an immersion, f, of a number of solid tori into $\mathbb{R}^4$.

$$f : \bigsqcup_i S^1 \times D^2 \to \mathbb{R}^4.$$

The disks $\{t\} \times D^2, t \in S^1$ are called the **fibres** of the immersion. The fibres are disjoint except at the points corresponding to real crossings. The restriction,

$$f \mid \bigsqcup_i S^1 \times \partial(D^2),$$

is called a **welded tube** and defines a number of knotted "ribbon" toroidal surfaces in $\mathbb{R}^4$ corresponding to welded knots.

Exercise 4.3. *Show that the behaviour of the welded tubes in R^4 is consistent with the allowable welded R moves.*

4.1.6 *Free Knots* (F, v)

The motivation for the definition of free knots is the following. Suppose we look at chord diagrams stripped of all ornamentation (no arrow heads, no signs) and apply all the changes suggested by the Reidemeister moves. Does this define a non-trivial theory?

The corresponding plane diagrams have free crossings tagged by F and virtual crossings tagged as usual by v. The R moves of the virtual crossings are as usual and v dominates F. The R moves of F are the same as the flat tag f with one extra move. This is because when the planar diagram is being realized from a chord diagram we don't know whether a specific arc approaching a free crossing from left to right does so from the south east or the south west. This ambiguity is overcome by the **virtualization move** considered earlier.

Fig. 4.15.   The virtualization move.

In Fig. 4.15, on the left hand side, the thicker line appears to come from the south west to the crossing tagged F. After the virtualization move, it appears to come from the north west, allowing for the ambiguity.

The virtualization move is equivalent to an R_4 move, $R_4(v, F)$. We can see this by introducing two virtual crossings on one side, say, by an R_2 move and then using virtualization to remove two other virtual crossings.

Fig. 4.16.   The $R_4(F, v)$ move.

The smallest non-trivial free knot (in terms of free crossings) is the **3-morning star**, introduced by Manturov. This has six free crossings and four virtual crossings.

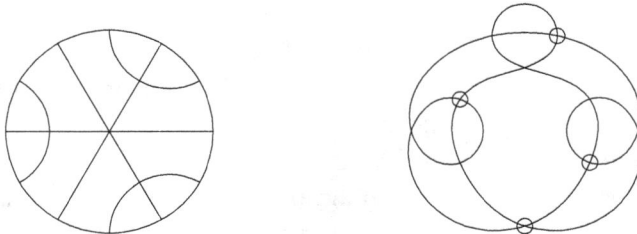

Fig. 4.17.   The 3-morning star as a chord diagram and as a plane diagram.

Chord diagrams can also be defined by a permutation. Let $X_1, X_2, \ldots, X_n$ be n points from left to right on the upper half of the boundary circle and let $Y_1, Y_2, \ldots, Y_n$ be n points from left to right on the lower half. If σ is a permutation of the numbers $\{1, 2, \ldots, n\}$ join X_i to $Y_{\sigma(i)}$ by a chord, $i = 1, 2, \ldots, n$.

For example, let $\sigma = (6125)(3874) \in S_8$.

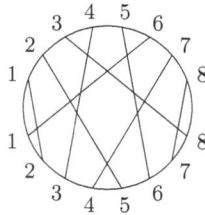

Fig. 4.18. A chord diagram defined by the permutation $(6125)(3874)$.

Both examples are minimal with respect to chords (free crossings) as we shall see when we come to consider invariants in a later volume.

4.1.7 *Virtual knots as knots on surfaces*

We now come to the main item of Chapter 4. In the previous chapter, we met virtual crossings as a solution to the realisation of chord diagrams which couldn't come, say, from a real knot. Remarkably, they also have an interpretation as curves on an orientable surface.

Suppose we have a virtual planar theory, $(\mathcal{A}, v)$, where none of the tags in $\mathcal{A}$ are virtual. We can convert this to a non-planar theory, $((\mathcal{A}))$, by eliminating the virtual tags. One way is described in Fig. 4.19. This introduces a handle on the plane and allows the arc to cross over the handle and eliminate a virtual crossing at the cost of placing the curve on a surface of higher genus.

Fig. 4.19. One way of eliminating a virtual crossing.

However, we will firstly use the following method. Consider a diagram, D, in $(\mathcal{A}, v)$ representing a knot $k \in (\mathcal{K}, v)$. The immersed curve of D is a graph and can be thickened to make a planar surface neighbourhood with boundary circles corresponding to the regions of the original diagram. At each virtual crossing lift the neighbourhood as shown in Fig. 4.20.

Fig. 4.20.   Another way of eliminating a virtual crossing.

The areas around the other vertices tagged by elements of $\mathcal{A}$ are left alone.

Fig. 4.21.   What happens to a non-virtual crossing, $a \neq v$.

The result is an oriented surface with boundary, Σ_0. Fill each boundary circle in with a disk. We now have a closed oriented surface, Σ, containing an immersed curve whose crossings are tagged by elements of $\mathcal{A}$. Write the resulting knot in $((\mathcal{K}))$ as $\Psi(k)$.

Theorem 4.3.   *The function* $\Psi : (\mathcal{K}, v) \to ((\mathcal{K}))$ *defined above is a bijection of knot theories.*

Proof.   The first step is to show that the map Ψ is well defined. In other words, the R moves in the plane do not affect the knot type in the target. We implicitly use the result that any homeomorphism of the surface with holes Σ_0 extends to a homeomorphism of Σ.

Consider firstly R moves which only involve $\mathcal{K}$ which means that no virtual crossings are used. All these moves need the support of a disk, be it a monogon, bigon or trigon. But by the filling in construction of Ψ, these disks will all be available. It follows that we only need to check those R moves which involve virtual crossings.

The action of $R_1(v)$ in the plane introduces or deletes a monogon. The effect on Σ_0 is to give a ribbon band an apparent twist, but nothing is really changed.

The move $R_2(v)$ introduces or deletes a bigon. This apparently seems to move a band of Σ_0 over another.

$R_3(v, v, x), x \neq v$ apparently moves a band over an x crossing. $R_3(v, v, v)$ apparently moves three bands about. In both cases, nothing much is changed.

The result of $R_4(x, v)$ on Σ_0 can be replicated by cutting it across four bands, two on one side, two on the other, reflecting horizontally and regluing the ends.

This shows that Ψ is well defined. In order to show it is a bijection, we need an inverse map Φ from $((\mathcal{K}))$ to $(\mathcal{K}, v)$. This is the inverse of the action indicated in Figs. 4.20 and 4.21.

In order to define this inverse map, we need to consider the following description of the surface Σ_g with g handles. On a 2-sphere which we could think of as a plane with a point at infinity, consider g disjoint squares. Cut out the interior of the squares and glue opposite edges by translations. The joined edges now form the meridian and longitude circles of the handles. To define a map $\phi : \Sigma_g \to S^n$, project the meridian and longitude circles onto the plane so that they become two intervals which cross in the middle.

Suppose we have a diagram, K, on the surface Σ_g. By a small adjustment, we can assume that the edges of K cross the meridian and longitude circles transversely. Consider a meridian and longitude pair in which m and n edges of K cross, respectively. After projection under ϕ, we will have created a square lattice the vertices of which now become mn virtual crossings. This defines the inverse map Φ from $((\mathcal{K}))$ to $(\mathcal{K}, v)$.

In order to show that this is well defined and an inverse to $\Psi : (\mathcal{K}, v) \to ((\mathcal{K}))$, we need to check three aspects.

(1) Any handle which doesn't involve K can be ignored as it doesn't effect Φ.

(2) Any R move can be performed clear of the handles and so takes place in the plane after ϕ. Clearly, this is an inverse procedure.

(3) Now, consider an orientation preserving homeomorphism $h :$ $\Sigma_g \rightarrow \Sigma_g$. This takes the diagram K to another diagram L say on Σ_g. Let K' and L' be the projected diagrams under ϕ in the plane. We must show that K' and L' are related by R moves.

To do this, we will use a result of Lickorish that every orientation preserving homeomorphism of a surface to itself is isotopic to a product of **Dehn twists**, see the paper by Lickorish [173]. Let α be a simple curve on an oriented surface, Σ. The Dehn twist about α is defined as follows. Consider an annular neighbourhood of α with two boundary circles. Cut along one of the boundaries, perform a 360-degree turn and then reglue. The action is illustrated in Fig. 4.22.

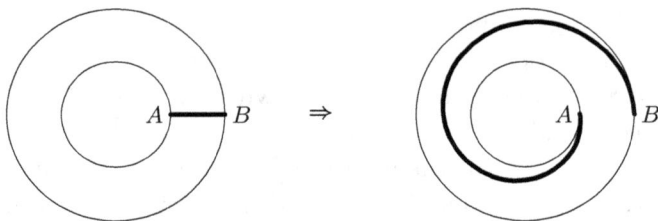

Fig. 4.22.   A Dehn twist.

Any transverse arc, such as the one from A to B, in Fig. 4.22 becomes a curled arc from A to B. We can assume that the intersection of the diagram with the annulus before the Dehn twist consists of a number of such transverse arcs. After the twist, these become disjoint curled arcs.

Now, push the surface into the plane. We can assume that the annulus is embedded after the push. If not, the annulus can be subdivided into subannuli for which this is true. The annulus will have originally run over a number of handles. After the push, we can assume that the annulus will intersect the diagram in a number of transverse arcs. These will intersect the curled arcs, which we label with the v tag. So, the curled arcs become virtual arcs, which we can then move them back to their original transverse arcs by a detour move.

Since any homeomorphism can be isotoped to a sequence of Dehn twists, we can deal with each in turn. $\qquad\square$

Chapter 5

Max Newman's Proof Technique

In the theory of rewriting systems, Newman's lemma, also commonly called the diamond lemma, states that a terminating (or strongly normalizing) abstract rewriting system (ARS), that is, one in which there are no infinite reduction sequences, is confluent if it is locally confluent.

— Wikipedia

5.1 Introduction

This chapter, as the name suggests, is a description of a marvellous technique introduced by Max Newman in 1942 for proving theorems not just in topology but throughout mathematics. So, this technique should be in all mathematicians' toolbox. Do not be spooked by the scary quote above. This method is simple in conception and application. If we could apply this technique to every unsolved mathematical problem, then its solution would drop into our hands like ripe peaches. Unfortunately, or perhaps fortunately, for the jobs of research mathematicians, this technique is not always available.

Before we consider the details, let's look at some examples. Consider Fig. 5.1. This is a complicated diagram of a classical knot with 10 crossing points. It may look complicated, but it happens to be the trivial knot!

Fig. 5.1.   The unknot in camouflage.

If you try to unknot this diagram using R moves, you will find that the number of crossing points increase before they decrease to zero.

This diagram, called **the culprit** by Kaufmann, is an example of a **hard unknot**, see the work of Kauffman and Lambropoulou [137].

Figure 5.2 is even more disturbing.

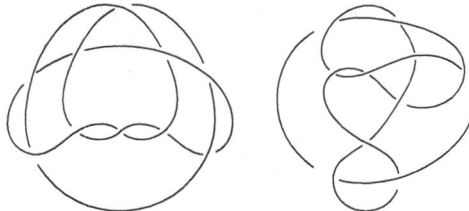

Fig. 5.2.   The Perko pair, $10_{161}, 10_{162}$.

The knot diagrams in Fig. 4.2. are clearly distinct and yet they represent the same knot. Both diagrams have the same minimum number, 10, of crossing points. We can get from one to the other by Reidemeister moves, but in doing so, we will have to increase the number of crossing points.

Mathematicians have tried to classify classical knots according to their **crossing number**. So, if K is a knot, then $c(K)$ is the minimum number of crossings for a diagram that represents that knot. In the late 1860s, the Scottish physicist William Thomson (Lord Kelvin) suggested that atoms were knotted vortices in the aether. If only we could better understand knots, we could unravel the secrets of the atom and of matter itself! Inspired by this theory, Thomson's countryman and fellow physicist, Peter Guthrie Tait, embarked on an investigation of knots which included the production of the first knot tables up to seven crossings.

Tait had no topological methods to distinguish his knots and had to rely on *ad hoc* methods. However, in 1926, Alexander and Briggs produced a bigger table, distinguishing the entries using techniques developed by Alexander. Further progress was obtained by Conway and Rolfsen used this method to produce the table in his famous book [236]. However, Perko, in 1974, found that two of the knot diagrams in the table represented the same knot and these are illustrated in Fig. 5.2.

The latest knot tables have nearly two million entries and the number rises exponentially as the crossing number increases. It is unlikely that any classification of classical knots by diagrams and crossing numbers will succeed because of the examples above. There is a local minimum which is not the absolute minimum, and even at the absolute minimum, there is no unique member.

Now, consider this very simple example. The words $u = aa^{-1}$ and $v = bb^{-1}$ both represent the trivial element, 1, in the free group on a, b. We can get from u to v in two ways:

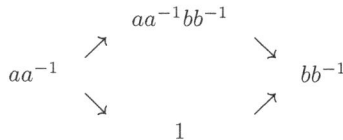

$$
\begin{array}{ccccc}
 & & aa^{-1}bb^{-1} & & \\
 & \nearrow & & \searrow & \\
aa^{-1} & & & & bb^{-1} \\
 & \searrow & & \nearrow & \\
 & & 1 & &
\end{array}
$$

The high road of this journey expands u to uv and then contracts to v. The low road contracts u to 1, the identity, and then expands to v. It should be pretty clear that the low road is the most efficient.

These examples illustrate situations where the methods of Newman work and where they do not, and we can sketch the outlines as follows. Recall that a *mathematical diagram theory* has objects represented by a diagram, although the word *diagram* in this context does not necessarily mean a geometric object. Diagrams are related by well-defined reversible moves and are said to represent the same object in the theory if there is a sequence of such moves between them.

If the diagrams are now assigned an order defined by some *height function* which means the moves either increase or decrease the height, then we can find a unique minimal representative provided the following two properties hold:

(1) *Any increase in height followed by a decrease can be reversed so that the decrease comes first followed by the increase.* This is the *diamond condition* relating to the lemma mentioned in the quotation at the beginning of the chapter.

(2) *There is no bottomless pit.* So, a downward sequence of moves must eventually terminate. This is the ARS mentioned in that quotation.

We have seen that Newman's technique does not give a unique diagram for a classical knot if the height function is the number of crossing points. It may be that some unknown height function may be discovered that does mean Newman's method can be applied to classical knots, but I personally doubt it.

On the other hand, elements of a free group are represented by words in the generators which take the role of diagrams. The length of one of these words is a suitable height function so that every element of a free group has a unique minimal length word to represent it.

5.2 Newman's classification graph

We will formalise the procedure sketched above. This procedure is common in mathematical proofs and was introduced by Newman [228]. It has since been recently considered by Bergman [18] and Bartholomew *et al.* [12].

Let G be a graph. We will denote the vertices of G by capital Roman letters $A, B, \ldots, K, L, \ldots$. If an edge joins K to L, we will write it as KL even though this may not define the edge uniquely.

A **path** in G from A to B is a sequence of vertices

$$A = K_0, K_1, K_2, \ldots, K_{n-1}, \quad K_n = B$$

in which K_{i-1}, K_i are joined by an edge, $i = 1, \ldots, n$.

A path is called **simple** if it has no re-entrant vertices, $K_i = K_j$, where $0 < i < j < n$. Any path can be replaced by a simple path with the same end points.

The graph is said to have **levels** if associated to any vertex K is an element of a partially ordered set, S. We write this association as $K \to h(K)$ and call $h(K)$ the **height** or **level** of K.

Recall that a partially ordered set has a binary operation $k < \ell$ satisfying the following axioms:

(1) Only one of the conditions $k < \ell, \ell < k, k = \ell$ can be true.
(2) If $k < \ell < m$ then $k < m$.

As usual, $h \leq \ell$ means either $h < \ell$ or $h = \ell$.

If $h(K) < h(L)$, we write $K < L$ and say K/L is **lower/higher** than L/K. Note that $h(K) = h(L)$ does not imply that $K = L$.

If an edge e of G has vertices K and L and $K > L$, then e is oriented from K to L. We will write an edge oriented in this manner as $e = K \searchbox L$ or $L \nearrow K$ and picture K on the page as being above L. We say that K **collapses** to L or L **expands** to K.

A path of the form

$$K \searrow K_1 \searrow K_2 \cdots \searrow K_{n-1} \searrow L$$

is called a **(monotonic) descending** path. The inverse of a descending path is called an **(monotonic) ascending** path.

We can now define an ordering on some pairs of vertices by defining $K > L$ if there is a descending path from K to L. This satisfies the transitive property. So, if $K > L > M$, then $K > M$.

We do not know yet if it is possible that $K > K$. But this will be excluded by the next condition.

A graph G with levels is said to have the **finite descending path property (FDPP)** if there are no infinite descending paths (see *ARS* in the quotation). More precisely, this means that if

$$s = K_1 > K_2 > \cdots$$

is any sequence, then there is an integer n such that

$$s = K_1 > K_2 > \cdots > K_n$$

In other words, the sequence always stops after a finite time.

Lemma 5.1. *Using the notation above, assume the finite descending path property. Then only one of the three conditions*

$$K = L, \quad K > L, \quad L > K$$

can be true.

Proof. Suppose $K > K$ is true. Then there is an infinite descending path

$$K > K > K \cdots$$

which is a contradiction. If $K > L$ and $L > K$ are both true, then $K > L > K$ implies $K > K$ by the transitivity property. A similar argument applies if $K = L$ and $L > K$ are both true.   □

A vertex R is said to be a **root** of K if there is a descending path from K to R which cannot be extended further. A root of G is a sink, so R has no outgoing edges. Therefore, $R \searrow L$ implies $L = R$. A root is a local minimum.

Lemma 5.2. *If a graph with levels has the FDPP property, then every vertex is either a root or is connected to a root by a descending path.*

Proof. If K is not a root, then it has a descending edge, $K \searrow K_1$. If K_1 is not a root, it has a descending edge, $K_1 \searrow K_2$, and so on. By the FDPP property, this process must terminate after a finite number of steps with a root R.   □

From now on, we will assume that all graphs considered have levels and satisfy the FDPP. Our task will be to discover which conditions are sufficient to guarantee that not only roots exist but also that they are unique. Consider the situation described at the beginning of the chapter. Let the vertices of a graph be classical knot diagrams and let the level of a vertex be the number of crossings. Then this graph has the FDPP, but we have seen that the roots are not unique.

We will call G a **strong Newman graph (SNG)** if the following two properties hold:

(1) G has the **Slanting Edge Property (SEP)** if for each pair of vertices, K, L, of G which are the end points of an edge, either $K < L$ or $L < K$.

(2) G satisfies the **Diamond Condition (DC)**. To define the diamond condition, we need a few definitions.

A **peak (valley)** is an ascending (descending) path composed with a descending (ascending) path. A peak (valley) is **simple** if it consists of just two edges.

Diamond Condition (DC): Any peak can be replaced by a valley with the same end points. In particular, if U descends to X and Y, then either $X = Y$ or there is a vertex V which ascends to X and Y.

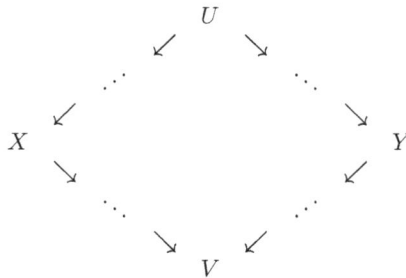

The diamond condition

So, a strong Newman graph has SEP and DC as above.

Lemma 5.3. *Every vertex of a path component of a strong Newman graph has the same unique root. Conversely, if a graph has the SEP and the property that every vertex of a path component has a unique root, then it is a strong Newman graph.*

Proof. Suppose the graph is a strong Newman graph and X descends to different roots R_1 and R_2. Then unless $R_1 = R_2$, this contradicts the DC. So, every vertex descends to a unique root.

Now, suppose that X and Y are joined by a path

$$X = K_0, K_1, K_2, \ldots, K_{n-1}, \quad K_n = Y$$

and yet X and Y descend to different roots R_1 and R_2. Then somewhere in the path, vertices K_i and K_{i+1} descend to different roots. We may as well assume that $K_i \searrow K_{i+1}$. Then K_i descends to one root and via K_{i+1} to another, contradicting the above.

Conversely, suppose a vertex U descends to X and Y. Then X and Y are clearly in the same path component and so both descend to a common root, V say. This is the diamond condition and so the graph is strong Newman. $\qquad\square$

We can localise the diamond condition as follows.

LDC: A graph has the **local diamond condition** if given a simple peak $X \nearrow U \searrow Y$, there is a path,

$$X = K_0, K_1, K_2, \ldots, K_{n-1}, \quad K_n = Y,$$

from X to Y such that if the path contains a simple peak $K_{i-1} \nearrow K_i \searrow K_{i+1}$, then there is an edge $U \searrow K_i$.

Lemma 5.4. *The local diamond condition (LDC) and the diamond condition (DC) are equivalent for graphs with the FDPP.*

Proof. Clearly DC implies LDC because a given a a simple peak $X \nearrow U \searrow Y$, there exists a valley by the DC and a valley does not have a peak. So, the LDC is satisfied vacuously.

Now, consider a graph with the LDC. Due to FDPP, every vertex which is not a root is connected to at least one root by a descending path. Our task is to show that this root is unique.

Let us call a vertex **regular** if it is connected to a unique root by a descending path: otherwise, we call it **irregular**. Clearly, regular vertices exist. A root is an example. Our task is to show that irregular vertices do not exist. We will assume the contrary and obtain a contradiction.

If an irregular vertex existed, then it clearly cannot be a root, and so in that case, there must be an irregular vertex, L say, such that for every edge $L \searrow K$, the vertex K is regular. If not, we could construct an infinite descending path of irregular vertices. Indeed, we will chose L so that every descending path from L must consist of regular vertices apart from L.

For such an irregular vertex L, there is a simple peak $X \nearrow L \searrow Y$ such that X and Y descend to unique but different roots. Chose X and Y so that the path, $X = K_0, K_1, K_2, \ldots, K_{n-1}, K_n = Y$ predicted by the LDC has shortest possible length.

The hypothesis of the LDC means that if the path joining X to Y has a simple peak $K_{i-1} \nearrow K_i \searrow K_{i+1}$ as a subpath, then there is an edge $L \searrow K_i$. This means that K_i is regular and so descends to a unique root.

This root must be different from one of the distinct roots of the pair X, Y. So, either the pair X, K_i or the pair K_i, Y has a shorter path.

So, the path joining X to Y has no simple peaks and is therefore either (a) ascending, (b) descending or (c) a valley. If (a) or (b), then X and Y have a common root. If (c) and if V is the base of this valley, then V has a unique root which must be the same for X and Y.

Therefore, irregular vertices cannot exist and all vertices are regular and connected to a unique root by a descending path. Hence, by Lemma 5.3, the graph is strong Newman and has the diamond condition. $\qquad\square$

We can now state a useful consequence of the above. Let

$$X = K_0, K_1, K_2, \ldots, K_{n-1}, \quad K_n = Y$$

be a path from X to Y. Then we write $h(K_0, K_1, K_2, \ldots, K_{n-1}, K_n)$ for the maximum value of the height $h(K_i)$, $i = 0, 1, \ldots, n$.

Lemma 5.5. *Suppose* $X \nearrow Z \searrow Y$ *is a simple peak in a strong Newman graph. Then there is a path* $X = K_0, K_1, K_2, \ldots, K_{n-1}, K_n = Y$, *from* X *to* Y *such that*

$$h(K_0, K_1, K_2, \ldots, K_{n-1}, K_n) < h(Z)$$

Proof. By hypothesis, there is a path $X = K_0, K_1, K_2, \ldots, K_{n-1}, K_n = Y$, from X to Y, satisfying the conditions of the hypothesis. If the path does not contain a simple peak, then

$h(K_0, K_1, K_2, \ldots, K_{n-1}, K_n) < h(X) < h(Z)$. If the path does contain a simple peak, then there is an edge $Z \searrow K_i$ for some i. So, $h(K_i) < h(Z)$. But this is true for all i, and hence, the result follows.

$\square$

The conclusion of Lemma 5.5 is a useful concept called the **peak condition**.

We now consider examples of strong Newman graphs.

Free Groups: The vertices of a graph are words in the symbols $x \in X$ and $x^{-1} \in X^{-1}$. The level of a word is its length. The edges due to expansions are insertions of pairs xx^{-1} or $x^{-1}x$ in the words. It is easy to see that this is a strong Newman graph. Hence, every word is equivalent to a unique reduced word with smallest length.

Planar Doodle Graph: Recall that plane doodles have a planar diagram with only one type of crossing tagged by d, where $\bar{d} = d$ and only the moves $R_1(d)$ and $R_2(d)$ are allowed. The vertices of the planar doodle graph are irreducible planar doodle diagrams and the edges are defined by the two allowable R moves, namely R_1 and R_2. The level or height is the crossing number of the diagram.

Theorem 5.1. *The planar doodle graph defined above is a strong Newman graph.*

Proof. The R moves always change the level, so the graph satisfies the slanted edge property, SEP. Now, consider a simple peak. This will consist of an expansive R_1 or R_2 move followed by a contractive one: there are four possibilities.

Each move creates/deletes a region Ω. If the two regions are disjoint, then we can interchange the moves and create a simple valley.

If the two regions coincide, then one is the inverse of the other and the simple peak can be replaced by equality.

So, we can assume that the above two situations do not happen. This leaves us with the possibilities illustrated in Fig. 5.3.

Part (a) of the figure shows the two types of regions: either a bigon or a mongon. Part (b) assumes that the two regions have one crossing point in common. Part (c) assumes that they have two crossing points in common.

It is easy to see that the situation defined by the left side of (b) consisting of two R_2 moves can be replaced by an equality with one local crossing and the situation on the right can be replaced by an equality with one or no local crossings. Either way, the number of crossings is reduced.

Part (c) assumes the existence of floating circles which is not possible if the doodle is irreducible.

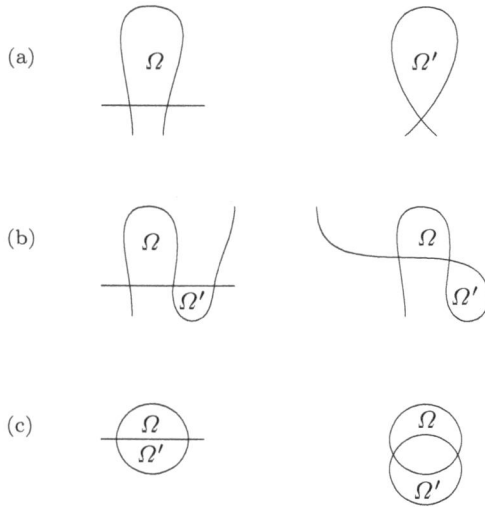

Fig. 5.3. Situations (a), (b) and (c).

It follows that every doodle has a unique diagram with the least number of crossings. This diagram will have no monogons or bigons.

5.3 (Weak) Newman graphs

We can expand the applications by generalising the form of graphs as follows. The vertices still have a height $h(K)$ defined by a partial order, but now, it is possible for the vertices of an edge to have the same height and we write $K \xrightarrow{0} L$ or $K \to \overset{0}{\cdots} \to L$ for a path which doesn't change its level. We call such a path **horizontal**. Recall that the maximum level in a path is denoted by $h(K \to \cdots \to L)$.

The conditions we need are as follows:

The Peak Condition (PC): Suppose $K \nearrow L \searrow M$ is a simple peak. Then there is a path $K \to \cdots \to M$ with $h(K \to \cdots \to M) < h(L)$.

The Transport Condition (TC)[9]: Let $K \nearrow L \overset{0}{\to} M$ be a path of length two, the first edge of which raises and the second has constant level. Then there is a path $K \to L_1 \to \cdots \to L_q \nearrow M$ such that $h(K \to L_1 \cdots \to L_q) < h(L) = h(M)$.

A **(weak) Newman graph** is a graph which satisfies the above.

Theorem 5.2. *Suppose R_1 and R_2 are roots of vertices K_1 and K_2, respectively, in the same component of a Newman graph. Then R_1 and R_2 can be joined by a horizontal path.*

Proof. Let $R_1 \to \cdots \to R_2$ be a path joining the roots. If it is not horizontal, it contains a subpath which is a plateau,

$$\nearrow M_1 \to \overset{0}{\cdots} \to M_q \searrow$$

By repeated application of TC, we can assume that the plateau is a simple peak. Now, an application of PC reduces the total height. Eventually, the path becomes horizontal. $\qquad\square$

We can now apply Theorem 5.3 to virtual knots. Let K be a knot diagram on a surface of genus g representing a knot in $(\mathcal{K})$. If g is minimal, we call it the **ambient genus** of the knot represented by K and write $g = \mathrm{ag}(K)$.

Theorem 5.3. *Consider two knot diagrams representing the same irreducible knot in $(\mathcal{K})$ and both on surfaces of minimal genus. Then the diagrams are equivalent under the R moves of $\mathcal{K}$ and homeomorphisms of the underlying surface.*

Proof. Consider the graph with vertices, diagrams of $(\mathcal{K})$ and edges defined by R moves, homeomorphisms of the underlying surface and null surgeries. Let the level be the genus of the underlying surface.

---

[9]It is called so because a level increasing move is *transported* to the right.

Only null surgeries change the genus, so the conclusion of the theorem is that there is a horizontal path joining the two diagrams. This will follow if we can show that the graph is Newman.

We first look at the peak condition, $K \nearrow L \searrow M$. This consists of adding a handle and then deleting one. By an isotopy, we can assume that the feet of the handles are disjoint. Since the diagram is not affected and the handles are disjoint from the diagram, K and M are equivalent by a homeomorphism.

We now look at the transport condition $K \nearrow L \xrightarrow{0} M$. If the belt b of the new handle is disjoint from M after $L \xrightarrow{0} M$, then we can interchange the operations possibly using a homeomorphism of the surface. Otherwise, suppose $L \xrightarrow{0} M$ is a positive R_1 or R_2 move which creates a disk, D crossing b. We can assume that the intersection of D with b consists of a number of disjoint arcs with end points on the boundary. Take an inner most arc spanning a sub disk D' whose boundary consists of the arc and part of the boundary of D and whose interior is disjoint from M. Now, push the arc off of the disk D over D' by an isotopy. Keep repeating until b is disjoint from M. □

It follows that if two knot diagrams representing the same knot both lie in the same surface of minimal genus, then they are related by homeomorphisms and R moves.

Another application of this idea is to show that the singular braid monoid embeds in a group, as shown by Fenn *et al.* [71]. Here, the vertices are singular braids. The expansions are the introduction of pairs of cancelling singular crossings. The levels are the number of singular crossings. We will look at this sort of argument in the following chapter.

Chapter 6

Generalised Braids

A braid is an intertwining of some number of strings attached to top and bottom "bars" such that each string never "turns back up." In other words, the path of each string in a braid could be traced out by a falling object if acted upon only by gravity and horizontal forces. A given braid may be assigned a symbol known as a braid word that uniquely identifies it (although equivalent braids may have more than one possible representations).

— Wolfram Mathworld

Classical braid groups were introduced explicitly by Emil Artin in 1925, although (as Wilhelm Magnus pointed out in 1974) they were already implicit in Adolf Hurwitz's work on monodromy from 1891.

Braid groups may be described by explicit presentations, as was shown by Emil Artin in 1947. Braid groups are also understood by a deeper mathematical interpretation: as the fundamental group of certain configuration spaces, as automorphisms of a free group and they are intimately associated with the Yang Baxter equation. As example of a classical braid is illustrated in Fig. 6.1.

Fig. 6.1. A classical braid represented by the word $\sigma_2\sigma_1\sigma_3^{-1}$.

In this chapter, we will generalise the classical braid concept and associate to each knot theory, $\mathcal{K}$, a braid theory consisting of a sequence of groups, $B_n(\mathcal{K}), n = 1, 2, \ldots$. These will be the generalised braids.

6.1 Braids from generalised knot theories

Let $\mathcal{K}$ be a (generalized) knot theory with tags $\mathcal{T} = \{a, \bar{a}, b, \bar{b} \ldots\}$ attached to crossings as usual. Let $I_n = \{1, 2, \ldots n\}$ be the first n integers, with the natural ordering. A **generalised n string braid representative of length** m is defined by a function

$$\alpha : I_m \rightarrow I_{n-1} \times \mathcal{T}.$$

A braid representative, α, can be made into a geometric **diagram** as follows. Most writers represent the braid vertically like the hair style. But we will represent them horizontally to respect their algebraic structure. Let X_m be a set of $m+2$-ordered points in the real line $\mathbb{R}$,

$$X_m = \{x_0, x_1, \ldots, x_m, x_{m+1} \mid x_0 < x_1 < \cdots < x_m < x_{m+1}\}$$

The braid diagram has n horizontal lines, $[x_0, x_{m+1}] \times I_n$-oriented from left to right and m vertical **bridges** joining the ith line $[x_0, x_{m+1}] \times \{i\}$ to the $(i + 1)$th line $[x_0, x_{m+1}] \times \{i + 1\}$, $i = 1, 2, \ldots, n - 1$ as follows. Suppose α is written as $\alpha(j) = (i_j, a)$, $j = 1, 2, \ldots, m$, $i_j \in I_{n-1}$ and $a \in \mathcal{T}$. Then there is a bridge, labelled by a, joining the point (x_j, i_j) to the point $(x_j, i_j + 1)$. Two such diagrams, corresponding to the same α, with different sets of m points, X_m and X'_m, are considered the same. Such diagrams are called **brick diagrams** of the braid.

An equivalent type of diagram is called a **crossing diagram** of the braid. Here, the bridges are replaced by crossings. This gives a more conventional view of a braid with strings which cross at double points. By following the strings from left to right, we can pick up an associated permutation of I_n. For instance, in Fig. 6.2, the braid induces the permutation $1 \to 2 \to 4 \to 3$.

However, both pictures are clearly interchangeable.

Both types of diagram are illustrated in the following example. Let $\mathcal{T} = \{a, b, \ldots\}$, $n = 4, m = 3$ and let α send $1, 2, 3$ to $(2, a), (1, \bar{b}), (3, b)$, respectively.

Fig. 6.2. Brick and crossing diagram example.

We have coloured the strings on the right of the diagram for emphasis.

6.2 The monoid structure

Braid representatives with the same string number multiply by concatenation in the usual way (see Fig. 6.3). The multiplication is associative since all ordered sets X_m are considered equivalent. There are identity elements with $m = 0$.

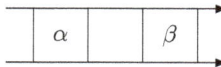

Fig. 6.3. The product $\alpha\beta$.

Braid diagrams, α, β can also be **stacked** like a column vector as in Fig. 6.4. The result is written as a row vector, (α, β).[10]

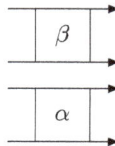

Fig. 6.4. Stacked braids (α, β).

---

[10] Another notation is $\alpha \otimes \beta$.

If α has p strings and β has q strings, then (α, β) has $p + q$ strings.

The subsets, X_m of the two braids can be intermeshed arbitrarily, which changes the diagram but not the braid because of **B$_1$** below.

Let $\mathcal{T}^+$ be the subset of the set of tags, $\mathcal{T}$, defined so that $a \in \mathcal{T}^+$ if $\bar{a} = a$ and if $\bar{a} \neq a$, then we chose one of $a, \bar{a}$ to be in $\mathcal{T}^+$, but not both. These define the **positive** generators of the monoid structure illustrated in Fig. 6.5, where $a \in \mathcal{T}^+$.

Let 1 denote the identity element and let a stand for a_1 in B_1. Then the generator a_i can be written in stacked form

$$a_i = (1, 1, \ldots, a, \ldots, 1)$$

with a in the ith position.

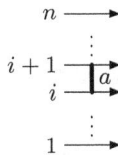

Fig. 6.5.   The generator a_i.

Historically, the generator for real braids is written as $r_i = \sigma_i$.

In the example of Fig. 6.2, the braid representative can be written as $a_2 \bar{b}_1 b_3$ in terms of the generators.

We now consider relations, **B$_1$, B$_3$, B$_4$**, illustrated in Figs. 6.6–6.8.

The commuting relation **B$_1$** illustrated in Fig. 6.6 is *always* obeyed.

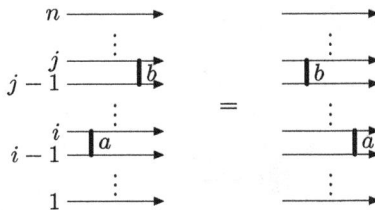

Fig. 6.6.   **B$_1$** : $a_i b_j = b_j a_i$ for all $|i - j| > 1$.

The next two relations depend on the allowable moves of the theory.

Suppose $R_3(a, c, b)$ is allowed. Then we have the following relations

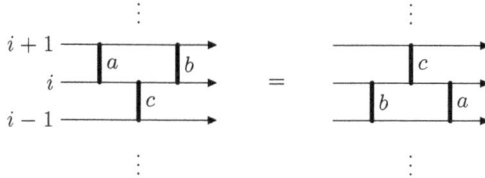

Fig. 6.7. $\mathbf{B_3} : a_i c_{i-1} b_i = b_{i-1} c_i a_{i-1}$, $i = 2, \ldots, n-1$.

If $R_4(a, b, c, d)$ is allowed, then we have the following relations.

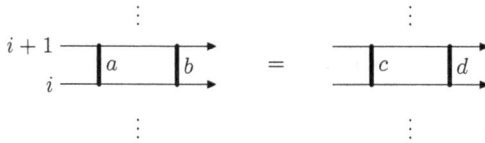

Fig. 6.8. $\mathbf{B_4} : a_i b_i = c_i d_i$, $i = 1, \ldots, n-1$.

6.3 Groupification

We can convert the monoid structure into a group, $B_n(\mathcal{K})$, the **generalised braid group on** $\mathcal{K}$, as follows. Define inverses by a new set of relations, $\mathbf{B_2} : a_i \bar{a}_i = \bar{a}_i a_i = 1$ for all i and all $a \in T^+$. Cancellation is shown in Fig. 6.9.

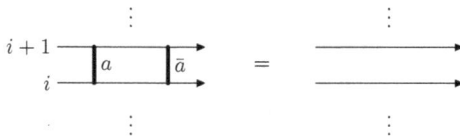

Fig. 6.9. The inverse element, $a_i^{-1} = \bar{a}_i$.

So, if $\bar{a} = a$, then $a_i^2 = 1$. If $\bar{a} \neq a$, then $a_i \neq a_i^{-1}$.

Let $B_n(\mathcal{K})$ denote the group obtained by adding inverses as above. For example, in Fig. 6.2, the braid represents the group element $a_2 b_1^{-1} b_3 = a_2 b_3 b_1^{-1}$.

In summary, if $\mathcal{K}$ is a generalized knot theory, there is a sequence of groups, $B_n(\mathcal{K}), n = 1, 2, \ldots$, with generators $a_i, b_i, \ldots, i = 1, \ldots, n-1$ corresponding to the tags, $a, b, \ldots$ of $\mathcal{K}$. These generators always satisfy a commuting relation, $a_i b_j = b_j a_i, |i - j| > 1$.[11] In addition, there maybe **extra relations** added according to the R moves satisfied by the theory.

6.4 Examples

The original braid groups, B_n,[12] corresponding to classical knots have generators σ_i and extra relations

$$\sigma_i \sigma_j \sigma_i = \sigma_j \sigma_i \sigma_j, |i - j| = 1$$

The **twin** groups, $TW_n = B_n(\mathcal{D})$, correspond to planar doodles. The generators d_i only have the extra relations $d_i^2 = 1$. They are right-angled Coxeter groups and have been studied in the work of Khovanou [85].

There are two more braid groups with one tag.

One is the group of symmetries, $S_n = B_n(\mathcal{F})$, with flat tag, f and extra relations $f_i^2 = 1$ and $f_i f_j f_i = f_j f_i f_j, |i - j| = 1$.

The other is $B_n(\mathcal{O})$ with tag o and no extra relations. They are the analogues of final and initial objects respectively in a category.

There are natural maps.

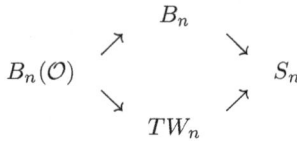

$$
\begin{array}{ccc}
 & B_n & \\
 \nearrow & & \searrow \\
B_n(\mathcal{O}) & & S_n \\
 \searrow & & \nearrow \\
 & TW_n &
\end{array}
$$

The kernel of the right-hand maps are called **pure** braids.

If there is more than one tag, then they are usually related by dominance. Recall that x dominates a, written $x \succ a$, if $R_3(x, x, a)$

---

[11] We usually drop these previous conditions as there are always $n - 1$ generators for each tab and the commuting relations are obligatory.

[12] We use the original notation, $\sigma_i = r_i$, and B_n for a classical braid generator and group.

is allowed. This can also be written as $R_3(x, x, \bar{a})$ if the theory is regular. Both imply the extra relation, $x_i x_{i-1} a_i = a_{i-1} x_i x_{i-1}$.

The singular braid group, $B_n(\mathcal{S})$, has two sorts of generators σ_i, s_i. The classical tags σ satisfy their usual relations. The dominance relations $\sigma \succ s$ and $\sigma^{-1} \succ s$ imply mixed extra relations, $\sigma_i \sigma_{i-1} s_i = s_{i-1} \sigma_i \sigma_{i-1}$ and $\sigma_{i-1} \sigma_i s_{i-1} = s_i \sigma_{i-1} \sigma_i$. Finally, there is the R_4 relation, $\sigma_i s_i = s_i \sigma_i$.

The virtual tag, v, defines the relations $v_i v_{i-1} a_i = a_{i-1} v_i v_{i-1}$ for all a_i as v dominates. Since $v_i^2 = 1$, it follows that $v_{i-1} v_i a_{i-1} = a_i v_{i-1} v_i$ for all a. One consequence to this is that every generator a_i is conjugate to every other generator a_j.

The weld tag, w, dominates but is dominated by the positive real tag, $r = \sigma$, but not by the negative real tag. So, $\sigma_{i-1} \sigma_i w_{i-1} \neq w_i \sigma_{i-1} \sigma_i$, but $\sigma_i \sigma_{i-1} w_i = w_{i-1} \sigma_i \sigma_{i-1}$, reflecting the two forbidden moves.

6.5 Embedding the monoid

A knot theory, $\mathcal{K}$, is called **positive** if the R_3 and R_4 conditions can be written without inverses. This means that the group relations for $B_n(\mathcal{K})$ can be written as monoid relations of the form $u = v$ where u and v are words in the positive generators $\mathcal{T}^+$. These relations define a monoid $M_n(\mathcal{K})$ associated to a positive theory. The usual theories such as classical, virtual, singular, and flat virtual are all positive.

Exercise 6.1. *Show that if $a \neq \bar{a}$, then the condition $R_3(a, \bar{a}, a)$ cannot be written as a positive relation in $B_n(\mathcal{K})$.*

There is a natural map, $\phi : M_n(\mathcal{K}) \to B_n(\mathcal{K})$ which is injective in certain cases, for example, the positive classical braid monoids embed, as shown in the work of Epstein *et al.* [254]. The same is true for singular braids, as shown by Fenn *et al.* [71].[13,14]

---

[13]There is a preprint which extends these results. If the argument is correct, then it will appear in a later edition.
[14]Care is needed with R_4. The condition $R_4(a, a, a, b)$ with $a \neq b$ prohibits injection.

We now prove another example. Recall that $\mathcal{O}$ is the theory of untagged diagrams, i.e. shadows with no R moves.

Theorem 6.1. *The monoid $M_n(\mathcal{O})$ embeds in its groupification $B_n(\mathcal{O})$.*

Proof. The monoid, $M_n(\mathcal{O})$, has a presentation:

$$< o_i | o_i o_j = o_j o_i, |i - j| > 1, i, j = 1, \ldots, n - 1 >$$

We form $B_n(\mathcal{O})$ from $M_n(\mathcal{O})$ by adding inverses, $o_i^{-1}, i = 1, 2, \ldots$ If α and β are monoid elements which become equal in $B_n(\mathcal{O})$, then as words in the generators they are related by a sequence of moves of two sorts; either insertions or deletions of two types of pairs $o_i^{-1}o_i$, $o_i o_i^{-1}$ or commutations of the sort $o_i^{\pm 1}o_j^{\pm 1} = o_j^{\pm 1}o_i^{\pm 1}, |i - j| > 1$. Clearly, commutation moves can be swapped with insertions or deletions.

Each insertion/deletion geometrically creates/kills a bigon with vertices tagged with o and $\bar{o}$. Let $\beta_1 \to \beta_2 \to \beta_3$ be an insertion followed by a deletion and let D_1 and D_2 be the created and killed bigons, respectively.

There are three cases: (1) $D_1 \cap D_2 = \emptyset$, (2) D_1 and D_2 have a vertex in common, or (3) $D_1 = D_2$. If (1), the two moves can be interchanged. If (2), β_3 is obtained from β_1 by swapping o with o^{-1}. If (3), then $\beta_1 = \beta_3$.

By standard arguments, we can eliminate all insertions or deletions, and α and β are equal in the monoid $M_n(\mathcal{O})$.   □

6.6 Some simple braid invariants

It is convenient to imagine that the strings of a braid are coloured by a colour initiated on the left of the braid. Suppose that C_i is the colour of the ith string. During the changes of the braid under the various allowed moves, this colour is constant.

Let us look at how the tagging changes during an R_3 move.

Note that along each of the three strings, the order of the two tags changes, the central triangle is rotated through 180 degrees

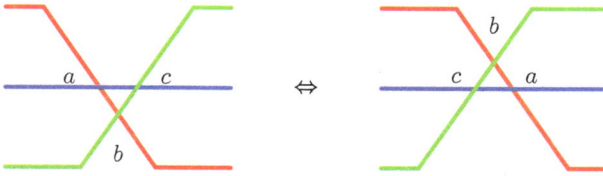

Fig. 6.10. Tagging changes.

about the centre of the edge labelled ac and the tag on the intersection of two strings is unchanged. We can use the above facts to make the following invariants.

For simplicity, assume there are no R_4 moves. Let A denote the abelian group generated by the tags, $\mathcal{T}$ and relations $\langle \bar{a} = -a \mid a \in \mathcal{T} \rangle$. For each $i = 1, \ldots, n$, let χ_i be the element of A defined by summing the tags along the ith string.

In the example braid of Fig. 6.2,

$$\chi_1 = -b, \quad \chi_2 = a + b, \quad \chi_3 = a - b, \quad \chi_4 = b$$

Note that these invariants are not independent.

We now go non-commutative and let G be the group with generators the tags, $\mathcal{T}$ and the relations $\langle \bar{a} = a^{-1} \mid a \in \mathcal{T} \rangle$. Let i and j be two integers in the range, $1 \leq i < j \leq n$. Define χ_{ij} by proceeding left to right along the i and j strings and noting the tags where they cross.

In the example braid of Fig. 6.2,

$$\chi_{12} = 1, \quad \chi_{13} = b^{-1}, \quad \chi_{14} = 1, \quad \chi_{23} = a, \quad \chi_{24} = b$$

We can extend χ_{ij} to all values by symmetry and by putting $\chi_{ii} = 1$.

Theorem 6.2. χ_i and χ_{ij} defined above are invariants of any braid which does not come from a theory with an R_4 move.

6.7 Braids and knots

In this chapter, we have shown how given a knot theory, $\mathcal{K}$, with tags T satisfying various R moves, there is a corresponding stratified braid group $B(\mathcal{K})$. It is possible to move in the opposite direction and associate a braid with a knot. From a braid β, we can make a diagram

$\hat{\beta}$ by the construction of closure. We can describe this process as follows. Imagine the braid as lying on the sphere, $S^2 = \mathbb{R}^2 \cup \infty$, along a ribbon neighbourhood of the equator. Now, join the right-hand ends of β to the left-hand ends by oriented parallel arcs in the neighbourhood. The result is a planar diagram, $\hat{\beta}$ as in Fig. 6.11.

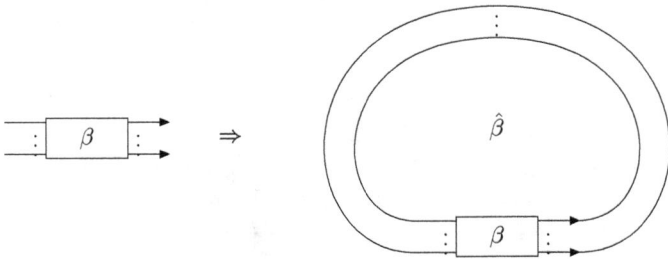

Fig. 6.11.   Closing a braid.

Exercise 6.2. *Show that a diagram of the trefoil knot can be obtained by closing the braid* σ_1^3.

Exercise 6.3. *Find a classical braid whose closure is the figure eight knot.*

Exercise 6.4. *Find a doodle braid whose closure is the Borromean rings. Repeat for the poppy.*

In the following chapter, we will investigate this process further. In particular, we will show when two different braids close to the same knot.

Chapter 7

Alexander and Markov Theorems

7.1 Introduction

In the previous chapter on braids, we demonstrated how a braid diagram from $B_n(\mathcal{K})$ can be converted into a knot diagram in the theory $\mathcal{K}$. We call such a diagram **braided**. The theorem that any (classical) knot can be represented by a braided diagram was first proved by Alexander [1] in 1923. Further result that any two braided diagrams representing the same knot can be joined by a sequence of Reidemeister moves in which the intervening diagrams are themselves braided was proved by Markov [218], in 1935.

Since then there have been several reproofs of these results by many others [18, 168, 224, 256, 264, 270]. These all relate to classical knots. A paper on virtual and welded knots has been published by Kamada [101], and there is also a paper on doodles by Gotin [77].

In this chapter, we shall prove analogues of the Alexander and Markov results for classes of generalized knots defined earlier and called regular and normal, respectively, see the work of Bartholomew and Fenn [8]. Any knot theory which satisfies the second Reidemeister move is regular. The examples of theories which are normal and so satisfy the hypothesis of the Markov type theorem include classical, virtual, welded, singular, virtual doodles and others. The theory of planar doodles is regular but not normal. So, regular theories satisfy an Alexander-type theorem, and in a subsequent chapter, we will show how this chapter's methods can be modified to prove a Markov type theorem for doodles.

The proof of the generalised Alexander theorem given here satisfies what I believe is Hardy's definition of beautiful mathematics and is due to myself and Bartholomew. Furthermore, the idea of the proof comes from Vogel's proof in the classical case.

The proof of the generalised Markov theorem, also due to Fenn and Bartholomew could not be described as beautiful. Under the lash of Hardy's dictum that there is no permanent place for ugly mathematics, I have tried to make the proof as easy to read as possible. But it requires looking at various cases which is always a downer and the reader on first visit may want to skip the proof and just accept the result.

7.2 Seifert cycles, graphs and trees

In a knot diagram, we can *smooth* crossings in an echo of the brick representation of braids. Surround each crossing by an oblong neighbourhood called a **crossing bridge**, labelled by the same tag as the crossing they replace, as shown in Fig. 7.1.

Fig. 7.1. Smoothing a crossing.

The crossing bridge or bridge for short can now be drawn as a slightly thicker line. The diagram now becomes a **Seifert graph**, consisting of a number of disjoint oriented simple loops, which we will call **cycles**, together with the bridges which join some pairs of cycles. The cycles are normally called Seifert cycles, see the paper by Seifert [243].

Since a cycle is oriented, it is the boundary of a right-hand disk and a left-hand disk. A pair of distinct cycles are disjoint and divide the sphere into two discs and a separating annulus. If both cycles are oriented in the same direction, i.e. are homologous cycles in the bounding annulus, then they are called a **coherent** pair.

Otherwise, they are **incoherent**. Note that if two cycles are joined by a bridge, then they are necessarily coherent. The annulus between a coherent pair is the intersection of a right-hand and a left-hand disk of the pair.

Let $h = h(K)$ be the number of incoherent pairs of the Seifert graph obtained from the diagram K. If $h = 0$, the diagram is braided and it is isomorphic to a diagram in which the cycles are circles of latitude and the bridges are arcs of longitude joining cycles. In other words, it is the closure of a braid.

A pair of cycles s_1 and s_2 are said to be **adjacent** if there is a path in the sphere with end points in s_1 and s_2 and whose interior is disjoint from the diagram. Note that adjacent cycles may be coherent or incoherent.

The components of the complement of the Seifert graph are called **regions**. A region whose boundary is a single cycle is called a **polar region** and its boundary a **polar cycle**. Note that the interiors of polar regions are disjoint from the rest of the diagram if the diagram is connected. It is easy to see that in this case, the number of polar regions is at least 2 and is only 2 if K is braided.

If a point of origin, O, is taken in one of the two polar regions of a braided diagram, then we can assume after an isotopy that a vector from O to a variable point on K will wind monotonically about O in the same direction for each component of K.

Let K be a knot diagram. If the diagram is connected, we can define an associated oriented tree $T(K)$, see the paper by Vogel [264]. The edges are in bijective correspondence with the cycles and the vertices are defined as follows. Strip away all the bridges and there is a vertex for each component of the complement of the cycles. The connections are defined so that two edges share a vertex if the associated cycles are adjacent. A right-hand normal from a cycle defines the orientation of the corresponding edge of the resulting graph which is a tree since we are on a sphere or plane.

The n-**chain** tree, C_n, is an interval divided into n edges. In any tree, two edges e_1 and e_2 can be connected by a unique n-chain in which e_1 and e_2 are the end edges. Let s_1 and s_2 be cycles and let e_1 and e_2 be the corresponding edges in $T(K)$. If the orientation of

e_2 is the same as the orientation induced by e_1 along the chain, then the cycles s_1 and s_2 are coherent and conversely.

The following observation and subsequent discussions are needed for the generalised Alexander theorem.

Lemma 7.1. *Suppose K is a connected knot diagram on the 2-sphere. If K is braided, then the cycles, which are all oriented coherently, are totally ordered by inclusion of their right(left)-hand discs. In particular, the tree, $T(K)$, is a coherently oriented chain. If K is not braided, so $h(K) > 0$, then there are a pair of adjacent incoherent cycles.*

Proof. If K is braided, the number of polar regions is 2 and $T(K)$ is a chain. Since $h(K) = 0$, $T(K)$ is coherently oriented and provides a total order on the cycles of K.

If K is not braided, the number of polar regions is greater than 2 and $T(K)$ has at least three ends. Let e_1 and e_2 be the end edges of a chain in $T(K)$ which have opposite orientations. Then there will be edges e_1' and e_2' in $T(K)$ which have opposite orientations and share a vertex. The corresponding cycles will be adjacent and incoherent. $\square$

7.2.1 *A worked example*

A knot diagram is shown in Fig. 7.2 together with its Seifert graph and tree. The bridge types corresponding to the crossings are not indicated.

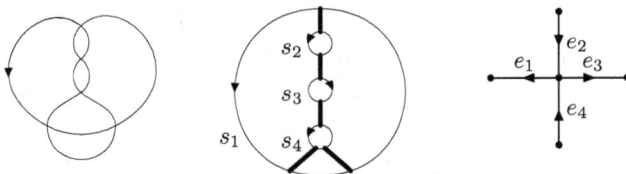

Fig. 7.2. Knot diagram K, its Seifert graph and tree $T(K)$.

Let s_1 be the outermost cycle, s_2 the top one of the inner cycles, s_3 the middle inner cycle and s_4 the bottom one of the inner cycles. The edges $e_1, \ldots, e_4$ in the tree correspond to these cycles.

There are five vertices corresponding to the open regions when the bridges and cycles are removed. The four 1-valent vertices correspond to the three small discs with boundaries s_2, s_3 and s_4 and the infinite disk with boundary s_1. The inner 4-valent vertex corresponds to the region with four boundaries.

The edge e_2 is oriented into the central vertex because a right-hand normal on the cycle s_2 would point away from its inner disk. The other edges are oriented in the same manner and the reader is encouraged to verify their orientations.

All four cycles are adjacent, s_1 is joined to s_2 and s_4 by bridges as is s_3. The cycle s_1 is coherently oriented with s_2 and s_4 but not s_3. The cycles s_2 and s_4 are coherently oriented with s_3, but s_2 and s_4 are not coherently oriented. So, $h = 2$. All cycles are polar.

7.3 The generalised Alexander theorem

An R_2 move creates/deletes a bigon. They are noted $R_2^{\pm}$ according to creation or deletion. The bigon is called a birth/death disk according to its existence or otherwise.

As we saw earlier, an R_2 move can be divided into two kinds. If the arcs are oriented together, say from left to right, then this is a parallel R_2 move and is denoted by R_2'. These preserve the value of h.

If the arcs are oriented in opposite directions, then this is a non-parallel R_2 move and is denoted by R_2''. These change the value of h and they are further divided into two types. If an R_2'' move involves the arcs from two distinct cycles, then this is called a **Vogel** or V move: V^+ if it creates birth disk, V^- otherwise. If the non-parallel R_2 move involves only one cycle, it is called a **non-Vogel move**. Vogel moves will play an important role in the subsequent proofs, as we can see from the following lemma.

Lemma 7.2. *A V^+ move which creates a bigon decreases h by 1.*

Proof. The idea of the proof is to show that if s, s' are two adjacent incoherent pairs of cycles, then the V move eliminates them and replaces them with a coherent pair. The details are as follows.

Assume that each cycle lies in the right disk of the other. Then the annulus between them is the intersection of their right-hand disks. Let l, a and r be the number of cycles incoherent with s which lie in the left-hand disk, the annulus and the right-hand disk, respectively. But do not include s' in the calculation for a and r. Define l', a' and r' similarly, but note that $a = a'$. Let $h(s)$ denote the number of cycles incoherently oriented with s and $h(s')$ the number of cycles incoherently oriented with s'. Then

$$h(s) = l + a + r + 1 \quad \text{and} \quad h(s') = l' + a + r' + 1$$

Now, do a V^+ move which eliminates s and s' and introduces coherent cycles c and c' joined by two bridges. One of these cycles is polar. Assume that c' is the polar cycle. Then

$$h(c) = l + l' + a \quad \text{and} \quad h(c') = r + r' + a$$

Since $2h = \Sigma h(s)$, summed over all cycles, it follows that h is reduced by 1. □

Exercise 7.1. *Show that the V^+ move in Lemma 7.2 has the effect on the Seifert tree, T, as shown in Fig. 7.3, namely the two incoherently oriented edges e_1 and e_2 are amalgamated into an edge e together with their end baggage in a scissor motion. At the same time, a new end edge f is created.*

Fig. 7.3.　The scissor move.

If the theory is regular, then R_2 moves are always allowed. Most importantly for regular theories, V moves are allowed.

Theorem 7.1 (The Generalised Alexander Theorem [11]).
In a regular knot theory, a series of positive Vogel moves, one for each value of h, can make all the Seifert cycles of a diagram

coherently oriented. So, h becomes 0, the Seifert cycles are nested and the diagram is braided.

Proof. Let K denote the diagram and suppose $h(K) > 0$. By Lemma 7.2, there is a pair of incoherently oriented cycles which are also adjacent. Push the arc of one along the path joining one to the other to make a V^+ move. This reduces h by 1. If h is still positive, continue this process until $h = 0$. $\qquad\square$

The shortness of the proof above belies its power and shows the utility of Seifert's idea. So, a complicated diagram with $h = n$ can be reduced to a braided diagram in n moves, provided the theory is regular.

However, if we consider untagged diagrams as forming a knot theory, $\mathcal{O}$, in which no R moves are allowed, then this theory is not regular and any diagram which is not braided cannot be made braided.

7.4 Orientation and the value of h

The R moves will change a diagram and possibly change h, the number of incoherent cycles. In this section, we look at how orientations for the R moves impact on h.

An R_1 move creates/deletes a monogon. A creative move is denoted R_1^+ and a deleting move is denoted by R_1^-. The new monogon is a new polar cycle and is coherently oriented with its parent cycle. The monogon is called a **birth/death** disk. The value of h is increased with the appearance of the monogon unless h is originally 0 and the new cycle lies in one of the two polar regions. This special R_1 move is called a **Markov move**.

Earlier, we noted that non-parallel R_2 moves changed h.

If the three arcs of an R_3 move are all oriented from left to right, then this is a parallel R_3 move. The other possibility for an R_3 move is that the central trigon is oriented. Such an R_3 move can be written in terms of parallel R_3 moves and V moves, as we saw in the introduction. So, since we are always dealing with regular theories, we can discount this type of R_3 move.

We will restrict ourselves to three types of the fourth move $R_4(a, b, c, d)$.

The first two of these are written $R_4(a, b), RR_4(a, b)$, and swap a, b. The second also inverts one of the tags. The arcs can be oriented arbitrarily and h is unchanged.

The second type is called an **exchange move** and is written $X(a, b) = R_4(a, \bar{a}, b, \bar{b})$. The effect is to replace $a, \bar{a}$ with $b, \bar{b}$. This is a combination of two non-parallel R_2 moves. The first move deletes an oriented bigon, labelled $a, \bar{a}$ and then the second recreates it, labelled as $b, \bar{b}$.

The exchange move does not alter h and only changes tags. Consequently, the reader may be excused for wondering at the relevance of this kind of R_4 move when it is in reality only a combination of R_2 moves. However, like the Markov move which is in disguise as an R_1 move, the exchange move's importance lies in the polar regions.

Consider the following example. Let $\alpha, \beta \in B_n(\mathcal{K})$ and $a, b \in \mathcal{T}$. Then the closure of $\alpha a_n \beta a_n^{-1}$ (which now lies in $B_{n+1}(\mathcal{K})$) is the same knot as the closure of $\alpha b_n \beta b_n^{-1}$ and a similar fact holds when a and b are at the base of the braid.

This move does not need to occur in classical braid theory, so it may come as news to some readers that it is needed in other theories. We will present conditions for when it is needed later.

We can now list all the moves in a regular theory which preserve h:

(1) parallel R_2 moves;
(2) parallel R_3 moves;
(3) Markov moves;
(4) R_4 moves.

We will for shortness call these R^o moves.

7.4.1 *Finger moves*

Finger moves come in two sorts, i and ii, and are illustrated in Fig. 7.4. An i finger move is applicable in a regular knot theory and consists of a series of R_2 moves in a line but may stop short of R moves enclosing a crossing.

A *ii* finger move is applicable in a normal theory with dominant tag x and consists of R_2 moves and an R_3 move using the dominant tag. This results in the crossing with tag a being enclosed.

The tags, $y_1, \ldots, y_q$, are arbitrary. The tags, x_1, x_2, x_3, x_4, take the values, x and $\bar{x}$, according to the orientations of the crossing arcs.

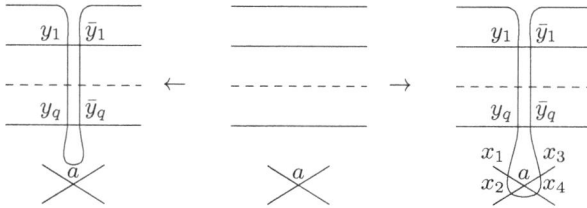

Fig. 7.4. *i* and *ii* finger moves.

7.5 R moves which change h and how we simplify them

The moves which change h are the following:

(1) R_1 moves when $h > 0$ and non-Markov R_1 moves when $h = 0$;
(2) non-parallel R_2 moves consisting of V moves and non-V moves.

In this section, we show that the non-Markov R_1 moves when $h = 0$ and non-parallel R_2 moves which are not V moves can be written in terms of (a) moves which do not change h (i.e. R^o moves) and (b) V moves. This simplifies the number of moves which have to be considered in the generalised Markov theorem.

In what follows, except in the proof of Lemma 7.3, we will leave off the tags in diagrams and assume that any R moves shown are allowed.

Lemma 7.3. *Let $r : K \to L$ be a positive R_1 move in a normal theory, which is non-Markov and suppose $h(K) = 0$. Then r can be written as a combination of V moves and moves which do not change h.*

Proof. Let the new crossing be tagged with a. Let x be a dominant tag.

Since h is raised, the birth disk must lie in an annulus defined by the parent cycle and one of the, say, $p > 0$ cycles between the birth disk and a polar region. Then $K \to L$ can be written as a i finger move sequence of p parallel R_2 moves tagged by $\bar{x}, x$ into the polar region. We then apply a Markov move tagged by a. Then p parallel R_3 moves using dominance and finally a negative i finger move of p inverse V moves. The combination is the non-Markov move illustrated in Fig. 7.5. □

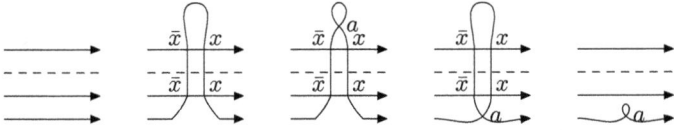

Fig. 7.5. Writing a non-Markov R_1 move in terms of other moves.

This is the first time that normality has been needed and is relevant when planar doodles are considered.

Lemma 7.4. *Let $r : K \to L$ be a positive non-parallel non-Vogel move in a regular theory. Then r is a combination of V moves, R_1 moves and moves which do not change h.*

Proof. Such a move only involves one cycle. Figure 7.6 shows how the move $K \to L$ can be rewritten as a sequence of 2 R_1 moves, a positive V move and a negative V move. The up or down arrows indicate whether or not h increases or decreases. □

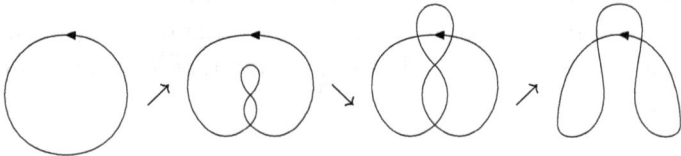

Fig. 7.6. Writing a positive non-Vogel move in terms of other moves.

In conclusion, the only moves we need to consider, which raise h, are the following:

(1) R_1 moves when $h > 0$;
(2) V moves.

7.6 Markov's theorem

In this section, we will prove the result that if two braided diagrams represent the same knot, then there is a sequence of R^o moves from one to the other, keeping the intermediate diagrams braided. The R^o moves needed are the ones defined earlier provided that the theory is normal. For non-normal theories such as planar doodles, a new move is needed and we will discuss this later.

The idea behind this proof is again to use Vogel moves to lower the values of h, but their application will be more dynamic than in the proof of the generalized Alexander theorem since the elements in a sequence of R moves will change during the application.

We can now state the theorem which generalises the classical Markov theorem.

Theorem 7.2 (Generalised Markov). *Suppose K and L are braided and define the same knot in a normal theory. Then they are related by a sequence of R moves which preserve braiding.*

Firstly, we will need some notation to clarify the statement of the theorem. A single R move will be denoted by $K \to L$. A sequence of R moves such as $K = K_1 \to K_2 \to \cdots \to K_{q-1} \to K_q = L$ of length q will be denoted by $K \to \cdots \to L$. If the moves do not alter $h(K)$, we write $K \overset{0}{\to} L$ or $K \to \overset{0}{\cdots} \to L$ and refer to them as **h-neutral** moves. The moves on braids which do not change h were listed earlier and consist of braid group changes, Markov moves and exchange moves.

A single R move which raises (lowers) h is denoted by $K \nearrow L$ ($K \searrow L$). The maximum value of h in a sequence is denoted by $h(K \to \cdots \to L) = \max(h(K_i)), i = 1, \ldots, q$. A sequence $K \nearrow L \searrow M$ is called a **simple peak** and a sequence $K \searrow L \nearrow M$ is called a **simple valley**. In view of the previous section, we can assume that, in a normal theory, the only R moves which change h are the V moves and non-Markov R_1 moves.

The statement of the theorem can now be rephrased as follows. In a normal theory, if two braided knots K and L are related by a sequence $K \to \cdots \to L$, then they are related by a sequence

$K \to \overset{0}{\cdots} \to L$ of R^o moves, in which all the intermediate diagrams are braided. So, $h = 0$ throughout the sequence.

We will use the methods of Chapter 5 to prove the result. As in the proofs at the end of that chapter, we will construct a graph, G, which has the knot diagrams as vertices and the edges as allowed R moves. The levels will be the number h of incoherent cycles.

If G^o denotes the subgraph of G consisting of braided diagrams and R moves which do not change the level h, then Markov's theorem says that any path in G with endpoints in G^o can be replaced by a path in G^o with the same endpoints.

The proof will follow if we can show that the graph G is a (weak) Newman graph. So, there are two conditions to be satisfied.

Lemma 7.5 (The Peak Condition). *Suppose $K \nearrow L \searrow M$ is a peak in a normal theory. Then one of the following is true:*

(1) *$K \cong M$;*
(2) *there is a sequence $K \to L_1 \cdots \to L_q \to M$ with $h(K \to L_1 \cdots \to L_q \to M) < h(L)$.*

Proof. Remember that we are only considering V moves and non-Markov R_1 moves when changing h.

The proof is divided into a number of cases:

(i) both moves are V moves;
(ii) the increasing move is a V move and the decreasing move is an R_1 move;
(iii) the increasing move is an R_1 move and the decreasing move is an R_1 move;
(iv) the increasing move is an R_1 move and the decreasing move is a V move.

The cases are further subdivided by the number of cycles involved, which can be 2, 3 or 4.

In case (i), if the moves involve the same two arcs of a pair of cycles, then either $K \cong M$ or there is an exchange move $K \to M$ of the form $X(a, b)$ which preserves h. If the tracks of the moves are disjoint, then the moves can be interchanged and we have a simple valley which is outcome 2.

If there are three arcs of three cycles involved, then the peak can be illustrated diagrammatically as in Fig. 7.7.

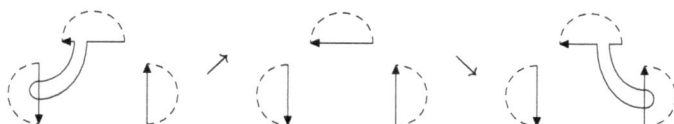

Fig. 7.7. A peak of V moves with three arcs.

We now replace the middle L by L' as in Fig. 7.8.

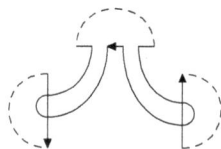

Fig. 7.8. L' the new L.

We now have a simple valley.

If there are four arcs of four cycles involved and the tracks of the V moves cross, then this can be illustrated diagrammatically as in Fig. 7.9.

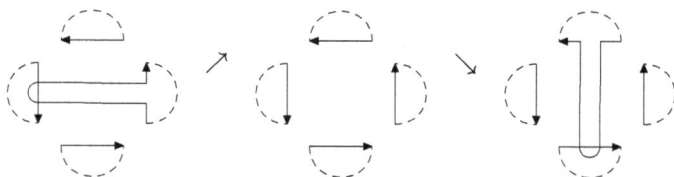

Fig. 7.9. The tracks of the V moves cross.

We now replace the middle L by L' as in Fig. 7.10.

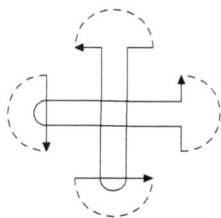

Fig. 7.10. L' the new L.

To get to L' from K involves one h neutral R_2 move and two positive V moves. So, h is lowered by 2. To get from L' to M involves two negative V moves and one h neutral R_2 move. The peak is therefore replaced by outcome 2.

Case (ii) is the reverse from right to left of case (iv) which we will consider in the following.

In case (iii), the birth and death disks are either equal or disjoint, which leads to outcome 1 or outcome 2, respectively.

In case (iv), if $h(K) = 0$, then this is covered by Lemma 7.3. If $h(K) > 0$ by Lemma 7.1, there is a positive V move r, say, whose track is disjoint from the birth disk and which lowers h by 1. If we start with r, then do the peak and then r^{-1}, and we will have outcome 2.   □

Lemma 7.6 (The Transport Lemma). *Let $K \nearrow L \xrightarrow{0} M$ be a sequence of two R moves, the first of which raises and the second keeps h constant. Then one of the following is true:*

(1) *the moves can be interchanged, $K \xrightarrow{0} L \nearrow M$;*
(2) *there is a sequence $K \to L_1 \cdots \to L_q \nearrow M$ such that $h(K \to L_1 \cdots \to L_q) < h(M)$;*
(3) *the sequence $K \nearrow L \xrightarrow{0} M$ has the following properties:*
    (a) *$h(K) = 0$;*
    (b) *$K \nearrow L$ is a V^- move which creates a new polar region P;*
    (c) *$L \xrightarrow{0} M$ is an R_4 move involving P.*

Proof. As with the peak lemma, there are several cases to consider. The increasing move may be a V^- move or, if $h(K) > 0$, an R_1 move which is not Markov. For each of these moves, there are three possibilities for the h-neutral move:

 (i) a parallel R_2 move;
 (ii) an R_3 move;
(iii) an R_4 move.

Assume initially that the first move is a V^- move. Then, if its track does not interfere with the second move, they can be interchanged; this will always be the situation in cases (ii) and (iii).

So, assume that the other move is a parallel R_2 move involving four cycles and that their paths cross. This is a similar situation to the one described in Fig. 7.10, but the direction of the top path is reversed (see Fig. 7.11).

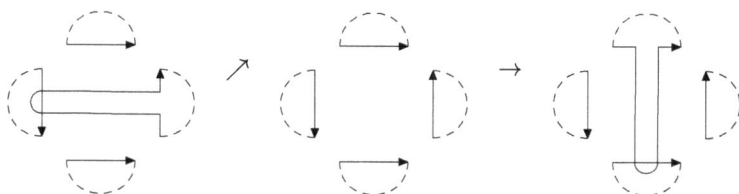

Fig. 7.11. A V^- move and an R_2 move which cross.

We can replace $K \nearrow L \to M$ by the sequence $K \searrow L_1 \nearrow L_2 \to L_3 \nearrow M$ as in Fig. 7.12.

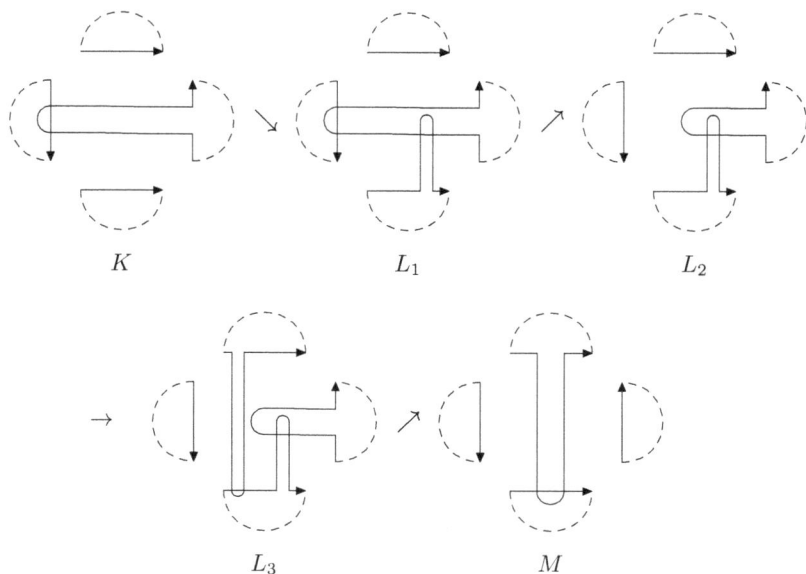

Fig. 7.12. $K \searrow L_1 \nearrow L_2 \to L_3 \nearrow M$.

The first two moves are V moves, the next is the required R_2' move and the last is another V move. Since $h(K \searrow L_1 \nearrow L_2 \to L_3) < h(M)$, we are in outcome 2.

Suppose now that $K \nearrow L$ is a positive R_1 move and $h(K) > 0$. As in the proof of Lemma 7.6, there is a positive V move r, say, whose track is disjoint from the regular neighbourhood and which lowers h by 1. As in the proof of the peak lemma, we can therefore start with r, do the R_2 moves and then do r^{-1} and we will have outcome 2.

Assume now that $h(K) = 0$ and $K \nearrow L$ is a V^- move which creates a new polar region P. Then the h-neutral move can be any of the three cases, (i), (ii), or (iii), listed above. The order of the moves can be swapped provided the track of the h-neutral move is disjoint from P, giving outcome 1. It is easy to see that this is the case if the h-neutral move is one of (i) or (ii), however for an R_4 move, it is possible for its track to be P. This is illustrated in the following diagram. In the left of Fig. 7.13, K, is braided and one of the polar regions is shaded. The other polar region is the external one. In the middle, a V^- move introduces a new polar region P, also shown shaded in L. The diagram on the right, M, is the result of an R_4 move with track P.

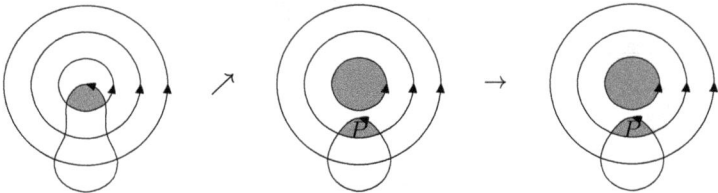

Fig. 7.13. $K \nearrow L \overset{0}{\to} M$.

This is outcome 3 and concludes the proof.   □

Proof of the generalized Markov. Suppose the braided diagrams K and L are related by a sequence $K \to \cdots \to L$. Consider a subsequence plateau $K' \nearrow K'' \to \overset{0}{\cdots} \to L'' \searrow L'$, where $h(K'' \to \overset{0}{\cdots} \to L'')$ is maximal and $h(K') > 0$. By repeated applications of the transport lemma, we can either eliminate the plateau or move the h raising move to the right until a peak is formed. Then the peak lemma allows us to reduce the value of h.

Assume now that $K' = K, L' = L$, so $h(K') = h(L') = 0$. Then, since we are considering only V moves and R_1 moves with $h > 0$ when changing h, we can assume there is a sequence $K \nearrow K' \overset{0}{\to} K'' \overset{0}{\cdots} \to L'' \overset{0}{\to} L' \searrow L$, where $K \nearrow K'$ and $L' \searrow L$ are V moves. Moreover, by the transport lemma, case 3, we may assume that $K' \overset{0}{\to} K''$ is an R_4 move with track P, say, and $L'' \overset{0}{\to} L'$ is an R_4 move with track P', say, and the regions P and P' are created by these V moves. Any other h-neutral moves are parallel R_2 moves, R_3 moves or R_4 moves, none of which interfere with P or P'.

However, in order to reduce h to zero, it is necessary to eliminate both P and P' by V moves. It follows that $P = P'$ and the final V move is inverse to the initial V move and both can be removed.

By combining all the arguments above, all the values of h are reduced to zero and the theorem is proved. $\square$

Chapter 8

Unfinished Business from Chapter 7

8.1 Introduction

There are a number of loose ends left over from Chapter 7 which need tidying up. The details come from the work of Bartholomew and Fenn [11]. In the previous chapter, we talked about braided diagrams. However, this could have been achieved without even considering braids. We know that a braid diagram, β, can be converted into a knot diagram, $\hat{\beta}$, by tying up the ends of the braid. Any diagram obtained in this manner is called **braided** and this process is described at the end of Chapter 6.

However, the braided diagram $\hat{\beta}$ does not completely define the braid β. Moreover, we need to consider how the choice of braid changes during the sequence of braided diagrams from $\hat{\beta}$ to $\hat{\beta}'$, representing the same knot as predicted by the generalised Markov theorem.

In addition, this Markov-type theorem does not work for planar doodles because the theory is not normal. So, anther kind of move is needed and this will addressed later in the chapter.

8.2 The Markov moves

The R^o moves considered in the previous chapter which preserve $h = 0$ can be divided into three types, M_1, M_2, M_3, by their actions on the braids which define the braided diagrams.

8.2.1 M_1: *Conjugation*

If a knot diagram is braided, then it has the form $\hat{\beta}$, where β is some braid in $B_n(\mathcal{K})$ say. We chose β by chopping the diagram, transversely avoiding the crossings. It is easy to see that the following simple lemma is true up to braid group moves.

Lemma 8.1. *Two braided diagrams, $\hat{\alpha}$ and $\hat{\beta}$, are equal if and only if α and β are related by the simple move*

$$\alpha = \alpha_1\alpha_2 \to \alpha_2\alpha_1 = \beta$$

for some decomposition $\alpha_1\alpha_2$ of α.

Moreover, it is clear that $\hat{\alpha}$ and $\hat{\gamma}$, $\gamma = \beta^{-1}\alpha\beta$, where $\beta \in B_n(\mathcal{K})$ represent the same knot by backdoor cancellation.

8.2.2 M_2: *Stabilisation*

Conjugation doesn't change the number of strings in a braid. Stabilisation changes the number of strings by one. We can embed an n string braid $\alpha \in B_n(\mathcal{K})$ into $B_{n+1}(\mathcal{K})$ by using one of the two ways:

$$\alpha \to (\alpha, 1)$$

introduces a new straight string at the top of α and

$$\alpha \to (1, \alpha)$$

introduces a new straight string at the bottom of α.

The following **stabilisation move** depending on a tag a can occur either at the top or bottom of the braid:

$$\alpha \to (\alpha, 1)a_n \quad \text{or} \quad (1, \alpha)a_1$$

It is easy to see that an $R_1(a)$ move proves that either move does not alter the knot type of the closure.

8.2.3 M₃: *Exchange*

The **exchange move** can be written as

$$(\alpha, 1)a_n(\beta, 1)a_n^{-1} \leftrightarrow (\alpha, 1)b_n(\beta, 1)b_n^{-1}$$

for an above move or

$$(1, \alpha)a_1(1, \beta)a_1^{-1} \leftrightarrow (1, \alpha)b_1(1, \beta)b_1^{-1}$$

for a below move. The braids α and β are in $B(\mathcal{K})_n$ and a and b are tags.

The first exchange move is illustrated in Fig. 8.1.

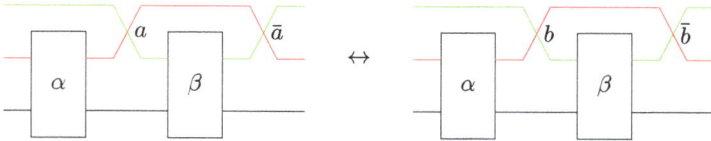

Fig. 8.1.   The exchange move $X(a,b)$ above.

Parts of the red and green arcs form a polar cycle when the braids are closed. The exchange of a and b is achieved by a $V^-(a)$ move and then a $V^+(b)$ move. This makes it clear that the knot type is unchanged.

Readers are invited to draw a diagram illustrating an $X(a,b)$ below move.

We will call the three moves above **M moves** and number them M_1, M_2, M_3 in order.

8.2.4 *Is the exchange move, M₃, really necessary?*

We will not answer this question completely. For real knots, the exchange move which interchanges r and $\bar{r}$ can be written in terms of conjugation and stabilisation, see the work of Traczyk [256]. A similar result is true for welded knots, where the interchange is between $\bar{r}$ and w, see the paper by Kamada [101]. These results are corollaries of Theorem 8.1.

Theorem 8.1. *Suppose $a, b,$ and c are tags in a knot theory $\mathcal{K}$ and $\alpha, \beta \in B_n(\mathcal{K})$. If the R_3 moves, (i) $R_3(b, c, \bar{b})$, (ii) $R_3(c, b, a)$, (iii) $R_3(\bar{b}, \bar{a}, c)$, and (iv) $R_3(a, c, \bar{a})$ are allowed in $\mathcal{K}$, then the exchange*

$$(\alpha, 1)a_n(\beta, 1)a_n^{-1} \leftrightarrow (\alpha, 1)b_n(\beta, 1)b_n^{-1}$$

is a consequence of the above as braid moves, conjugation and stabilisation.

If the R_3 moves, (i) $R_3(\bar{b}, c, b)$, (ii) $R_3(a, b, c)$, (iii) $R_3(c, \bar{a}, \bar{b})$, and (iv) $R_3(\bar{a}, c, a)$ are allowed in $\mathcal{K}$, then the exchange

$$(1, \alpha)a_1(1, \beta)a_1^{-1} \leftrightarrow (1, \alpha)b_1(1, \beta)b_1^{-1}$$

is a similar consequence.

Proof. Figure 8.2 shows the sequence of moves (down columns and left to right) performing the first exchange which is an above exchange.

For example, going from the top left to the diagram below, the blue arc performs an $R_2(b)$ move between a and $\bar{a}$. Then a Markov move involving c is performed. (By conjugation, a Markov move does not just have to take place at the side.) Then the next move down is an $R_3(b, c, \bar{b})$, which is allowed by hypothesis, and so on to the bottom and then top right and so on using the hypothesised moves.

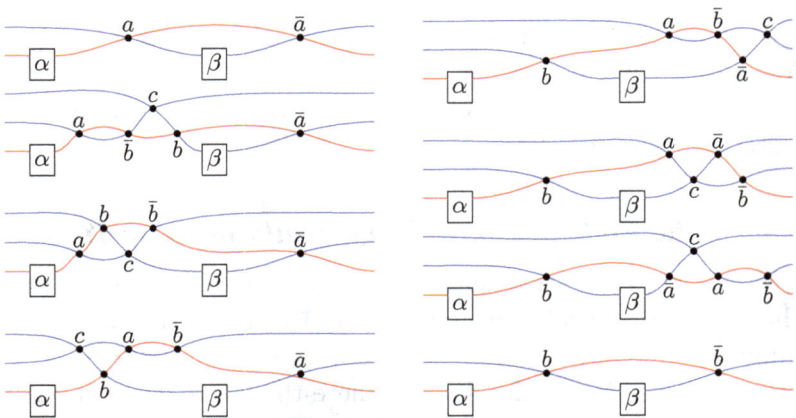

Fig. 8.2. The exchange move $X(a, b)$ above.

In a similar fashion, Fig. 8.3 performs the sequence of moves which initiates the lower exchange.

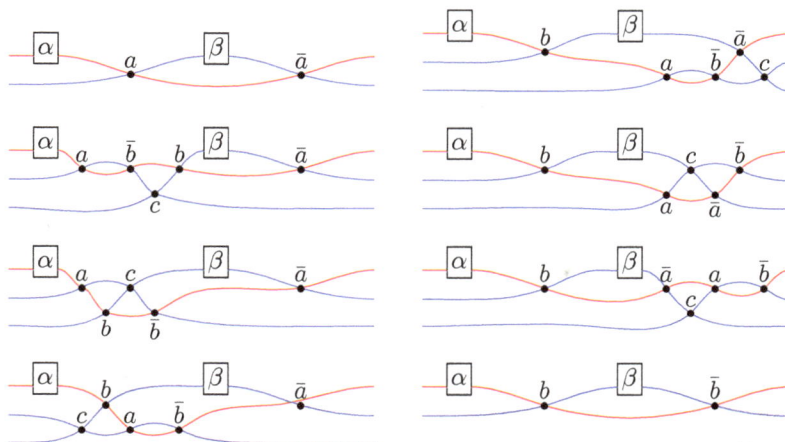

Fig. 8.3. The exchange move $X(a, b)$ above.

The description of the proof is similar to the above case.   □

The attentive reader will have noticed the symmetry between the above cases.

Corollary 8.1. *For real braids, the exchange $X(r, \bar{r})$ is not needed.*

Proof. Put $a = r, b = \bar{r}, c = \bar{r}$.   □

Corollary 8.2. *For welded braids, the exchange $X(\bar{r}, w)$ is not needed.*

Proof. Put $a = \bar{r}, b = w, c = w$.   □

As we remarked at the beginning of this section, this is not the whole story. We need an example and an invariant to show that virtual knot theory needs the exchange move, see the paper by Kamada [101]. This is beyond the scope of this book, but we will return to this in the next volume.

8.3 Doodles and twins

The twin groups TW_n are the braid groups associated with planar doodles,[15] $\mathcal{D}$. A presentation for TW_n has generators $d_1, d_2, \ldots, d_{n-1}$ and relations

$$d_i d_j = d_j d_i \qquad |i - j| > 1$$
$$d_i^2 = 1 \qquad i = 1, 2, \ldots, n - 1$$

The absence of a $\mathbf{B_3}$ relation reflects the fact that doodle diagrams are acted upon by R_1 and R_2 but not R_3 moves. Consequently, doodles have a unique minimal diagram with no monogons or bigons, as we saw in Chapter 5.

Since doodles are regular, they can all be represented as the closure of a twin. For example, the Borromean rings and the poppy are the closure of twins, $(d_1 d_2)^3$ and $(d_1 d_2)^4$.

However, a minimal diagram of a doodle is not necessarily braided as shown in the example in Fig. 8.4.

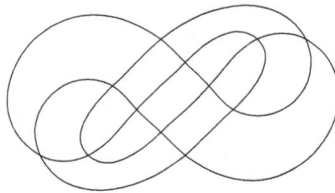

Fig. 8.4. A minimal doodle which is not braided.

This doodle is not braided because it has four polar regions.

Exercise 8.1. *Recall that a region whose boundary is a single cycle is called a* **polar region** *and its boundary a* **polar cycle**. *Orient Fig. 8.4 and find the four polar regions.*

Planar doodles are not normal and a Markov theorem involving just the three moves considered earlier is not possible as we shall now see and rectify.

---

[15]The concept of a doodle is due to Fenn and Taylor [70]. In their definition, component curves were simple and so lacked self-crossings. This more general definition is due to Khovanov [156].

8.3.1 *Some properties of twins*

Let $w \in TW_n$. By an abuse of notation, we will not distinguish the word in the generators $d_1, d_2, \ldots, d_{n-1}$ from the group element that it represents. We call i an **even index** if d_i occurs an even number of times in w and an **odd index** otherwise. Note that the number of odd indices is a group-invariant and taken mod 2, it is an additive group-invariant.

Let $TW_n^{(2)}$ denote the commutator subgroup of TW_n and consider the following easily proved result.

Lemma 8.2. *The following statements are equivalent:*

(1) $w \in TW_n^{(2)}$;
(2) w *is in the kernel of the abelianisation homomorphism;*
(3) *all indices are even.* □

Let $w \in TW_n$, and place the elements of w around the unit circle, S^1, in an anti-clockwise fashion. For all even i, divide the appearances of d_i into pairs and join by an arc of a circle which is orthogonal to S^1 (a hyperbolic line). Call this an i-**arc**. The intersection of an i-arc and a j-arc if it exists is called an (i, j)-**cross**. Let $\zeta(w)$ be the sum mod 2 of all the (i, j)-crosses where $i \neq j$.[16]

Lemma 8.3. *Let $w \in TW_n$, then the sign, $\zeta(w)$, has the following properties:*

(1) $\zeta(w)$ *does not depend on the choice of pairings;*
(2) $\zeta(w)$ *is a homomorphism when restricted to the commutator subgroup, $TW_n^{(2)}$;*
(3) $\zeta(w, w') = \zeta(w) + \zeta(w')$, $((w, w')$ *is the* **stack** *of w and w');*
(4) $\zeta(w)$ *is invariant under the moves M1-3.*

Proof. Let P_1, P_2, P_3, and P_4 be points on the circle where d_i occurs and let Q_1 and Q_2 be points where d_j is paired and $i \neq j$. The points Q_1 and Q_2 either separate none, one or two of the P points. Obviously, the first case is irrelevant and if, say, P_1 is separated from

---

[16]There is no doubt a sequence of such signs, $\zeta_i(w)$, with $\zeta(w) = \zeta_2(w)$, corresponding to the lower central series and Massey product, but we will not pursue this here.

the rest then there is an (i,j)-cross however the P points are paired. If, say, P_1 and P_2 are separated from P_3 and P_4, then on any pairing, there are either 0 or 2 (i,j)-crosses.

Since the pairings all give the same answer, it is easy to see that $\zeta(w)$ is well defined and 2, 3, and 4 follow easily. $\square$

Consider $w = (d_1 d_2)^2$, the sign of w is 1 and w closes to the trivial doodle, which is the closure of $w_0 = d_1$ having sign 0. So, you cannot get from w to w_0 by the classic Markov moves. However, if you embed TW_n in VTW_n, the braid group coming from virtual doodles, then you have the exchange moves because the theory of virtual doodles is normal. So,

$$w = d_1 d_2 d_1 d_2 \to d_1 v_2 d_1 v_2 = d_1 v_1 d_2 v_1 \to d_1 v_1 v_1 = d_1$$

8.4 Markov for planar doodles

In the previous chapter, we proved a Markov type theorem for normal knot theories. The obvious example of a knot theory which is not normal is the theory of planar doodles. However, we will show that, by introducing a new move on braided planar doodles which preserves braiding, an appropriate Markov type theorem can then be proved. The example in the previous section showed that this new move is necessary.

As in the earlier proof, the idea is to use Vogel moves to lower the values of h, but that proof depended on the theory being normal, which doodles are not. The proof has had to be adapted so that instead of moving from a braided diagram to a braided diagram by the two R moves (or M1-2) only available to us,[17] we have to use an extra move defined in the following.

We now need to look at the available R moves which we can use and the extra move just for plane doodles. This will require a number of definitions common to the earlier proof. We have included those for the convenience of the reader.

A circular component of a doodle which spans a disk whose interior contains no crossings is called a **floating circle**. A doodle which

---

[17]Since there is only one tag d with $d = \bar{d}$, there is no exchange move.

has a diagram in two or more pieces is called **reducible**. Otherwise, the doodle is **irreducible**. We need only consider irreducible doodles. Clearly, irreducible doodles do not have floating circles.

Let

$$s_n^i = t_n t_{n-1} \cdots t_{i+1} t_i t_{i+1} \cdots t_{n-1} t_n$$

and

$${s_n^i}' = t_1 t_2 \cdots t_{i-1} t_i t_{i-1} \cdots t_2 t_1.$$

Lemma 8.4. *Let* $\alpha \in TW_n$ *be a twin. Then the closures of* α, $(\alpha, 1)s_n^i$, *and* $(1, \alpha){s_n^i}'$ *are all the same doodle.*

Proof. To get from the closure of α to the closure of $(\alpha, 1)s_n^i$ requires $n - i$, V^- moves and one R_1^- move, see Fig. 6.13. This is similar for $(1, \alpha){s_n^i}'$. $\qquad\square$

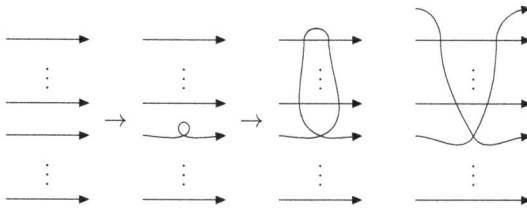

Fig. 8.5. The genesis of an augmented stabilisation move.

We will call the moves M_2^+: $\alpha \rightarrow (\alpha, 1)s_n^i$ and $\alpha \rightarrow (1, \alpha){s_n^i}'$ for any i **augmented stabilisation moves**.

Augmented because when $i = n$ or $i = 1$, respectively, these are the classical stabilisation moves on a classical braid.

Theorem 8.2. *Suppose two twins close to the same irreducible doodle. Then there is a sequence of* M_1 *(conjugation) and* M_2^+ *(augmented stabilisation) moves which takes one to the other.*

During this sequence, the closures of the twins will of course be braided. However, if $\hat{\alpha}$ and $\hat{\beta}$ are the same doodle, then there will be a sequence of doodle diagrams from $\hat{\alpha}$ to $\hat{\beta}$ related by R moves and these diagrams need not be braided. Our task is to make them braided by reducing h in each case to zero.

We will use the same notation and definitions defined earlier, but now the h-neutral moves will include the augmented stabilisation move. The peak and transport will need tweaking.

The proof of Theorem 8.2 will follow from the following two lemmas.

Lemma 8.5 (The Peak Condition for Doodles). *Suppose $K \nearrow L \searrow M$ is a peak in the doodle theory. Then one of the following is true:*

(1) $K \cong M$;
(2) *there is a sequence $K \to L_1 \cdots \to L_q \to M$ with $h(K \to L_1 \cdots \to L_q \to M) < h(L)$.*

Proof. We refer the reader to the earlier proof where all the details are the same except for cases (iii) and (iv) when $h(K) = 0$. Case (iii) is trivial, since for $h(K) = 0$, the birth and death disks must be equal, giving outcome 1. This just leaves case (iv) to consider further as follows. Suppose that $K \nearrow L$ is an R_1^+ move, $h(K) = 0$, say $K = \hat{\gamma}$, where $\gamma \in TW_n$, the R_1 move pushes up from the ith horizontal line of γ so that its birth disk lies between the ith and $i + 1$th horizontal line and $L \searrow M$ pushes the disk past the $i + 1$th horizontal line by a V^+ move. So, an augmented stabilisation move followed by a sequence of $n - i - 1$ V^- moves achieves the result. □

Lemma 8.6 (The Transport Condition for Doodles). *Let $K \nearrow L \overset{0}{\to} M$ be a sequence of two R moves, the first of which raises and the second keeps h constant. Then one of the following is true:*

(1) *the moves can be interchanged, $K \overset{0}{\to} L \nearrow M$;*
(2) *there is a sequence $K \to L_1 \cdots \to L_q \to M$ such that $h(K \to L_1 \cdots \to L_q) < h(M)$.*

Proof. Again, we refer the reader to the earlier proof where the details are the same except for the case $h(K) = 0$, $K \nearrow L$ is an R_1^+ move and $L \overset{0}{\to} M$ is an R_2' move that is not disjoint from the R_1^+ move. In this case, $K = \hat{\gamma}$ where $\gamma \in TW_n$, the R_1 move pushes up from the ith horizontal line of γ so that its birth disk lies beween the ith and $i+1$th horizontal line and $L \xrightarrow{0} M$ pushes the disk down past the ith horizontal line by an R'_2 move. Therefore, an augmented stabilisation move, a sequence of $n - i - 1$ V^- moves, the R'_2 move, and a final V^- move give outcome 2. $\square$

Proof of 5.12.10. The argument follows as before since the doodle graph is Newman. Suppose the braided diagrams K and L are related by a sequence $K \to \cdots \to L$. Consider a subsequence plateau $K' \nearrow K'' \to \overset{0}{\cdots} \to L'' \searrow L'$, where $h(K'' \to \overset{0}{\cdots} \to L'')$ is maximal. By repeated applications of the transport lemma, we can either eliminate the plateau or move the h, raising move to the right until a peak is formed. Then the peak lemma allows us to reduce the value of h. Eventually, all the values of h are reduced to zero and the theorem is proved. $\square$

Chapter 9

The Sorting Method: Invariants

9.1 Introduction

This chapter will look at easily defined invariants using combinatorial methods. Some invariants of generalised knots can be defined from classical analogues, but they tend not to be so useful. Powerful invariants having no counterparts in the classical case can be defined for generalised knots from generalised diagrams. This is only a quick trip through the subject because a complete treatment would easily fill a book. If I am spared, this is what I want to do.

If you want to distinguish two knots, you search for something which is defined for all knots and so is an *invariant*, but differs on the two knots in hand. Most knot theorists spend a lifetime looking for better and better invariants.

However, there is a trade off. Knots with the same invariant form clumps. The bigger the clumps, the weaker the invariant. For example, the 3-colouring invariant, described in the following, is fine for distinguishing the classical trefoil from the unknot but fails in the same task for the figure eight knot. On the other hand, the fundamental quandle can, in principle, distinguish all classical knots, but it is almost as difficult to classify as the knots it represents.

Nonetheless, there are advantages. Looking at an invariant such as a quandle changes the environment from geometry to algebra and algebraic techniques such as the representation can be used in the discrimination wars.

Invariants have two descriptions: geometric and combinatorial, with possible overlapping. The geometric description may be possible if the knot theory has a geometric description. A combinatorial description comes from the knot diagram, the fundamental concept in this book. The fundamental quandle has a geometric and combinatorial description, but the biquandle, a powerful invariant of virtual knots, is defined from a diagram, however a geometric definition using the description of a virtual knot as a curve in a thickened surface has so far eluded researchers.

The chapter is divided into sections which reflect which kind of invariant is defined.

9.2 Numerical invariants

The earliest invariant of classical knots was not 3-colouring but the diagram with the minimum number of crossings in any representative diagram of the knot. The Perko pair defined in Chapter 5 shows that this method is not foolproof. The early researchers had no other methods but a piece of string and their hands.

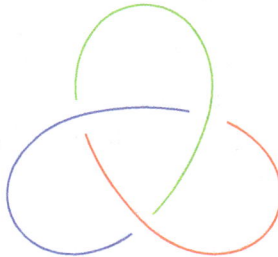

Fig. 9.1. The 3-coloured trefoil.

In Fig. 9.1, the classical trefoil knot has the continuous arcs coloured by three colours. The rules are simple: at each crossing, there are one or three colours but not two. This works for the classical trefoil but not the figure eight or indeed the virtual trefoil.

Exercise 9.1. *Show that 3-colouring really is an invariant by show-ing that the number of colourings by three colours is unchanged by the three R moves. Use the invariance to show that neither the Hopf link nor the classic Borromean rings can be pulled apart.*

Numerical invariants can mean integers, modulo numbers or ele-ments of other groups. These are often the easiest to understand or define but not necessarily the easiest to calculate.

We have already met some numerical invariants, namely the writhes and flat linking number of Chapter 3. Mostly they consist of finding some invariant set and counting the elements. We can illustrate this with the 3-colouring invariant. Consider the trefoil, by changing the colours of the arcs in Fig. 9.1, we get six possibil-ities. But there are also the three constant colourings making nine altogether. For the unknotted circle, there are only three possible colourings, so the trefoil knot is knotted. An alternative count is the number of colourings where at least two colours are used. For the trefoil, this is six for the unknot 0.

For another example, consider the Hopf link of two linked circles which can only have three constant colourings, whereas a pair of unlinked circles can have nine colourings, and so on.

9.2.1 *The maps S_a and T_a*

Three-colouring is an example of an edge colour or **EC-invariant**. Let us look at how this can be generalised. We can think of the edge of a knot diagram as a particle accelerator and a colour as a fundamental particle. As the two particles collide at a crossing, they split into two more particles. Specifically, let $\begin{pmatrix} x \\ y \end{pmatrix}$ be a column vector representing the incoming colours approaching a crossing tagged with a and let S_a be the function which changes the colours, so

$$S_a \begin{pmatrix} x \\ y \end{pmatrix} = \begin{pmatrix} x' \\ y' \end{pmatrix}$$

where $\begin{pmatrix} x' \\ y' \end{pmatrix}$ represents the outgoing colours. If X denotes the palette of possible colours, then S_a is a function, $S_a \colon X^2 \to X^2$, as illustrated in Fig. 9.2.

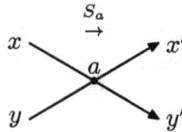

Fig. 9.2.   Action of the function S_a.

We call S_a the **switch map** associated to the tag a.

We now look at the conditions imposed on S_a by the moves R_1 to R_4.

R_1: If we look at Fig. 9.3, we see that in order to eliminate the monogon and using the above notation, we need the condition shown.

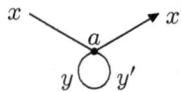

Fig. 9.3.   If $y' = y$, then $x' = x$.

If we now reflect Fig. 9.3 in a horizontal line, we get the reverse implication. Putting both together, we see that in order for $R_1(a)$ to be satisfied, we need the following condition:

$$y' = y \Leftrightarrow x' = x \qquad (*)$$

R_2: If we look at Fig. 9.4 which shows a parallel $R_2(a)$ situation, it is clear that S_a being inverse to $S_{\bar{a}}$ is necessary for $R_2(a)$.

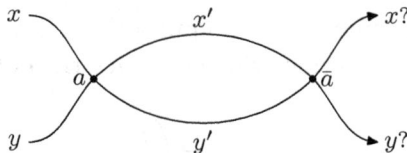

Fig. 9.4.   Why $S_a S_{\bar{a}} = id$ is necessary for $R_2(a)$.

Since we only deal with regular knot theory, we can always assume that S_a is a bijection.

Exercise 9.2. *Show that the conditions* $(*)$ *above and* $(**)$ *below can exist or not independently.*

$$S_a S_{\bar{a}} = \text{id} = S_{\bar{a}} S_a \qquad (**)$$

In other words, both conditions are necessary for S_a.

But this is not the end of the matter as we have to deal with the non-parallel $R_2(a)$. To manage this, we need a new map T_a called the **track map**[18] of the tag a.

We illustrate this in Fig. 9.5.

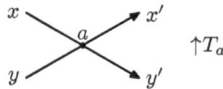

Fig. 9.5. Action of the function T_a.

We would like T_a to send the row (y, y') to the row (x, x'). But in order to do this, we need to define x uniquely given y'. This leads to the following condition:

For each $y \in X$, the function $X \rightarrow X$ defined by $x \rightarrow y'$ under S_a is a bijection.

Similarly, the following dual condition is required:

For each $x \in X$, the function $X \rightarrow X$ defined by $y \rightarrow x'$ under S_a is a bijection.

Exponential and lower fix notation for the maps, $x \rightarrow y'$ is $x \rightarrow x_y$ and $y \rightarrow x'$ is $y \rightarrow y^x$. Another notation used in the literature is $f_x(y)$ and $f^y(x)$, respectively, or even $y \wedge x$ and $x \vee y$, respectively.

---

[18]In many publications, this is called the *sideways map*, but that doesn't make sense here as the direction is upwards. In later publications this is called the twitch map.

So,

$$S_a(x, y) = (y^x, x_y) = (f_x(y), f^y(x)) = (y \wedge x, x \vee y)$$

These functions are like operations on the right acting on the left. Both operations are bijections and so have an inverse. We use the following notation for these inverses:

$$x^{y\bar{y}} = x^{\bar{y}y} = x = x_{y\bar{y}} = x_{\bar{y}y}$$

We have no need for brackets because y and $\bar{y}$ on the upper and lower levels are operators defined by y.

If we now look at Fig. 9.6, we see that the required condition for a non-parallel $R_2(a)$ is T_a and is inverse to $T_{\bar{a}}$.

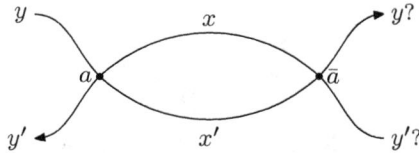

Fig. 9.6. Why $T_a T_{\bar{a}} = id$ is necessary for $R_2(a)$.

$\boldsymbol{R_3}$: If we look at Fig. 9.7 which shows the two sides of an $R_3(a, b, c)$ situation, it is clear that the compositions, $(S_a \times 1)(1 \times S_b)(S_c \times 1)$ and $(1 \times S_c)(S_b \times 1)(1 \times S_a)$, are either equal or different according to the rules of the knot theory.

Fig. 9.7. Both sides of $R_3(a, b, c)$.

$\boldsymbol{R_4}$: We won't consider R_4 in this chapter, but its use will necessitate equations such as

$$S_a S_b = S_c S_d$$

which we might deal with in another volume.

9.2.2 *The name of the game*

All these conditions have names and we will give a few here:

- If $a = b = c$ and conditions $R_2(a)$ and $R_3(a)$ are satisfied, then S_a and T_a are said to define the **birack** associated to the tag a.
- If in addition, the condition for $R_1(a)$ is satisfied, the birack is said to be a **biquandle**.
- If one of the maps is the identity, then we drop the prefix *bi-* and call the results **racks** and **quandles**, respectively.

Exercise 9.3. *Show that for a quandle and any x, either $x^x = x$ or $x_x = x$. Show that for a rack and any x, y, z, $x^{y^z} = x^{zy^z}$, (the self-distributive property).*

9.2.3 *Examples of biquandles: Back to 3-colouring*

The theory of biquandles and biracks is impressibly opaque at present and is a suitable topic for a future volume. So, we will stick to fairly simple examples here. Let $X = \{B, G, R\}$ be the three colours used to paint the edges of a knot diagram. Then they form a quandle under the rules

$$B^G = R, \quad R^B = G, \quad G^R = B$$
$$B^B = B, \quad R^R = R, \quad G^G = G$$

Exercise 9.4. *Show that for the 3-colour quandle,*

$$B^R = G, \quad R^G = B, \quad G^B = R.$$

We now generalise to n colours, $\{0, 1, 2, \ldots, n-1\}$ by the rule

$$i^j = 2j - i \bmod n$$

We will call this the n**-colouring quandle**.

Exercise 9.5. *Check that the n-colouring quandle is a quandle and that the operators are involutive.*

We can geometrically imagine the elements of the n-colouring quandle as the nth roots of unity $w_i = \exp(2i\pi/n), i = 0, 1, 2, \ldots, n-1$

on the unit circle. Then $w_i^{w_j}$ is the reflection of w_i in the line through the origin and w_j.

Consider Fig. 9.8. The figure eight knot is coloured for $n = 5$, albeit with only four colours. Note that the arc coloured by 4, changes its colour to -4 on passing under the arc coloured by 0. So $-4 = 1$ or $5 = 0$.

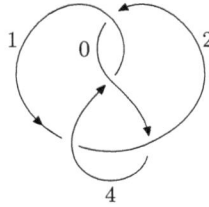

Fig. 9.8. Colouring the figure eight knot.

Since there is at least one non-constant colouring of the figure eight knot, it follows that the figure eight knot is non-trivial. Moreover, since the figure eight knot cannot be 3-coloured, it is not the trefoil.

Exercise 9.6. *Show that the figure eight knot cannot be n-coloured if n is not divisible by 5.*

9.2.4 *Invariants of virtual knots*

Almost all the invariants in the literature to date are for virtual knots. We will look at a trick due to Manturov [180] , which converts a quandle on a knot theory to a quandle on the virtualisation of the theory.

Consider Fig. 9.9: a crossing tagged by a is coloured by a quandle defined by the operation $y \to y^x$. On the right is a virtual crossing coloured with the biquandle defined by an invertible function q.

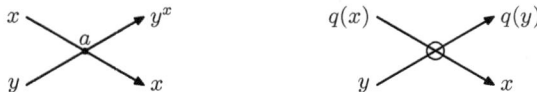

Fig. 9.9. A quandle and its virtualisation.

So, the S and T operators for the virtual tag v are given by

$$S_v \begin{pmatrix} x \\ y \end{pmatrix} = \begin{pmatrix} q(y) \\ q^{-1}(x) \end{pmatrix} \quad \text{and} \quad T_v(y, x) = (q(x), q(y))$$

Exercise 9.7. *Show that the function q above defines a biquandle for any tag a which satisfies $R_1(a), R_2(a)$ and $R_3(a, a, a)$.*

Exercise 9.8. *Show that S for the biquandle defined above is an involution and $T^{-1}(x, y) = (q^{-1}(y), q^{-1}(x))$.*

Consider a knot diagram which involves crossings with the tag a and virtual crossings. Suppose that associated to the tag a is the quandle x^y and associated to the virtual crossing is the biquandle defined by q.

Now, consider Fig. 9.10, which shows the virtual tag dominating the tag a with $R_3(v, v, a)$.

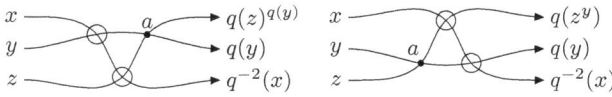

Fig. 9.10. $R_3(v, v, a)$.

In order to extend the colours, we need the **virtualisation** condition

$$q(x^y) = q(x)^{q(y)}$$

In other words, q is a **quandle automorphism.** Then the colouring defined by the left column of colours is preserved on both sides.

Exercise 9.9. *Let $q : X \to X$ be defined by $q(x) = x^{x_0}$ for some quandle and some fixed $x_0 \in X$. Show that q is a quandle automorphism.*

Now, consider Fig. 9.11 which shows the two sides of $R_3(a, a, v)$. Now, no matching of colours is possible unless the quandle action

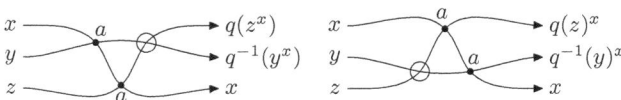

Fig. 9.11. $R_3(a, a, v)$.

defined by a satisfies $q(z^x) = q(z)^x$. This is possible if q is the identity map. This is the **weld situation** as shown by Fenn *et al.* [64]. Then the virtual crossing is replaced by a weld crossing, w, and a can now dominate w.

The reverse tag $\bar{a}$ implies a quandle action in the opposite direction, as indicated in Fig. 9.12.

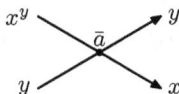

Fig. 9.12.   The reverse map.

If we now look at both sides of $R_3(\bar{a}, \bar{a}, w)$ in Fig. 9.13, we see that $\bar{a}$ does not dominate w unless the actions commute.

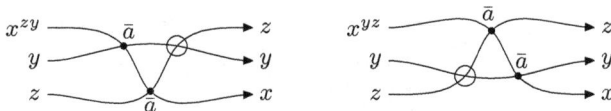

Fig. 9.13. $R_3(\bar{a}, \bar{a}, w)$.

In principle, this should lead to a proof that certain welded knots are non-trivial. In [8, 9], biquandles discovered by brute force computer work does the job.

Exercise 9.10. *If we replace the quandle in Fig. 9.10 by a biquandle, show that the colourings can be extended provided q is a **biquandle automorphism**, so $q(x^y) = q(x)^{q(y)}$ and $q(x_y) = q(x)_{q(y)}$.*

Exercise 9.11. *In the biquandle analogue of Fig. 9.11, find the conditions necessary on q for the colours to be extended.*

As an example, suppose the quandle is n-colouring. Since $w_i^{w_j}$ is the reflection of w_i in the line through the origin and w_j, the automorphism, q must preserve this. So, it is an element of the *dihedral group*, D_n, symmetries of the n-gon.

As an application, consider the virtual trefoil in Fig. 9.14. It is 5-coloured with q given by complex conjugation.

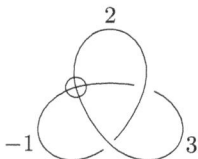

Fig. 9.14. Colouring the virtual trefoil.

It follows that the virtual trefoil is non-trivial.

9.3 Polynomials as invariants[19]

With a bit of extra work, we can find useful quandles and biquandles dependant on parameters, t, s, etc. Then some more work, perhaps with a computer, can extract invariant polynomials.

Consider the n-colouring quandle. This is an offspring of the **Alexander quandle** defined as follows. Let $\mathcal{A}$ be an affine space and t an invertible variable. If $\mathbf{a}, \mathbf{b} \in \mathcal{A}$, put

$$\mathbf{a}^{\mathbf{b}} = t\mathbf{a} + (1 - t)\mathbf{b}.$$

Geometrically $\mathbf{a}^{\mathbf{b}}$ lies on the line through $\mathbf{a}$ and $\mathbf{b}$ in the ratio $t : 1 - t$. If $t = -1$, this defines a reflection in $\mathbf{b}$. Of course, t doesn't have to be a real number. It could be any unit in a ring.

Exercise 9.12. *Show that the Alexander quandle is a quandle.*

Exercise 9.13. *Show that* $\mathbf{a}^{\bar{\mathbf{b}}} = t^{-1}\mathbf{a} + (1 - t^{-1})\mathbf{b}.$

If we combine the Alexander quandle with the function $q(\mathbf{a}) = s\mathbf{a}$ where s is an invertible variable, we have invented a very powerful invariant of virtual knots. This leads to a 2-variable polynomial which generalises the classical Alexander polynomial [8].

The method of getting the polynomial from a diagram using the Alexander quandle is described by Manturov and Ilyutko [217], but

---

[19]By *polynomial* in this context, we mean a Laurent polynomial.

we will also see how this is done in the following section from a skein relation.

Here, we are leaving this particular thread because the phrase *easily defined* is beginning to be betrayed. Readers who want more will have to wait for the second volume or read the literature in the bibliography.

9.4 Skein invariants

Take a crossing tagged by a in some diagram. Surround the crossing by a small disk centred at the crossing. We can now change the crossing in one of the ways illustrated in Fig. 9.15, but leaving the rest of the diagram unchanged.

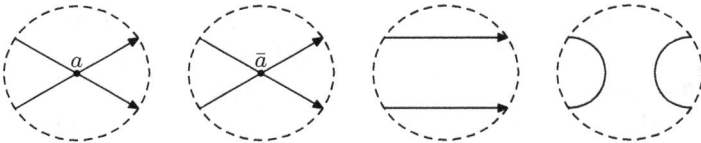

Fig. 9.15.   Three changes to a crossing tagged by a.

If K was the original diagram, then we notate the new diagrams by

$$K = K_+, K_-, K_0, K_\infty$$

A **skein relation** is a linear relation involving

$$\Psi(K_+), \Psi(K_-), \Psi(K_0), \Psi(K_\infty)$$

and an initial value of $\Psi(O)$ on the unknot, which defines Ψ as a value defined by the diagram. Hopefully, it will be unchanged under the R moves of some theory. For example, consider

$$\nabla(K_+) - \nabla(K_-) = z\nabla(K_0), \quad \nabla(O) = 1$$

where $a = r$, the classic crossing tag. In 1969, John Conway showed that this skein relation could be used to compute a version of the Alexander polynomial, now called the Alexander–Conway polynomial.

In 1984, Jones invented his own polynomial using the skein relation

$$t^{-1}J(K_+) - tJ(K_-) = (t^{1/2} - t^{-1/2})J(K_0), \quad J(O) = 1$$

Remarkably enough, these two A-list mathematicians, Conway and Jones, failed to notice that you could use two variables which would combine both definitions. This grandmaster blunder was soon put right by Hoste, Ocneanu, Millett, Freyd, Lickorish, Yetter, Przytycki and Traczyk who in a series of independent papers came up with the two variable HOMFLYPT polynomial bearing the first letters of their names.

One skein relation for the HOMFLYPT polynomial is

$$\lambda V(K_+) + \lambda^{-1}V(K_-) + \mu V(K_0) = 0, \quad V(O) = 1$$

It is not clear on first glance that the skein relation defines a knot invariant. One proof that it does relies on the fact that by changing the sign of the crossings, you can reduce any classical knot to the unknot. A more algebraic proof is given in an appendix.

Exercise 9.14. *Why isn't there a 3-variable HOMFLYPT polynomial?*

Exercise 9.15. *Calculate the HOMFLYPT polynomial of the following:*

(1) *the Hopf link;*
(2) *the right-hand trefoil knot;*
(3) *the left-hand trefoil knot.*

9.4.1 *The Kauffman bracket*

The **Kauffman bracket**, $\langle K \rangle$, is a polynomial in the variable A defined on *unoriented* classical knot diagrams by the skein relation

$$\langle K_+ \rangle = A\langle K_\infty \rangle + A^{-1}\langle K_0 \rangle, \langle O \rangle = 1$$

and

$$\langle K \sqcup O \rangle = (-A^{-2} - A^2)\langle K \rangle$$

A few words of explanation are needed here. Since we are dealing with unoriented diagrams, K_+ means rotate the crossing so that if it were locally oriented from left to right, then it would be a right-hand crossing. The second condition relating to a disjoint union of a diagram K and the unknot is necessary because the bracket so far only satisfies $R_2(r)$ and $R_3(r, r, r)$.

Exercise 9.16. *Show that the Kauffman bracket satisfies $R_2(r)$ and $R_3(r, r, r)$. Hint: There is a cunning proof in* [117].

Exercise 9.17. *Show that $L(K) = (-A)^{-3w(K)} \langle K \rangle$, where $w(K)$ is the writhe of the diagram, satisfies $R_1(r)$ in addition to $R_2(r)$ and $R_3(r, r, r)$.*

Now that we have a fully invariant definition, we can prove the following.

Theorem 9.1. *The polynomial $L(K)$ is equal to the Jones polynomial, $J(K)$, provided $A^{-2} = \sqrt{t}$*

Proof. Show that the polynomial in A satisfies the Jones skein relation if $t = A^{-4}$. □

9.5 Conclusion

This chapter has been a quick skim through some examples of knot invariants. There are plenty of edge labelling invariants of virtual knots but not enough skein types. There wasn't enough space to describe arrow polynomials or all the myriad representations of the fundamental quandle and biquandle. There is a book or even two books of work there.

The country of knot invariants has many pockets of civilisation. There are the large cities of bustling inhabitants studying quantum $SU(2)$ representations, Khovanov homology, Chern Simons invariants and other a-list enterprises. Outside of the cities in the rural areas are market towns of quandles, biquandles, and other invariants now abandoned by the elite but chewed over by the worthy burghers. Up in the foothills, along rugged overgrown lanes and vacated stone houses, a few inhabitants crack a precarious living from investigating stick numbers, genus, curvature, etc.

Appendix A

The Homflypt Polynomial

A.1 Introduction

In this appendix, we will give an algebraic account of the Homflypt polynomial, how it can be defined and its relation to the classic braid group. We will look at the Hecke algebra and trace functions in order to define the Homflypt polynomial.

A.2 Hecke algebra representations

In this section, we will study representations of the braid groups into Hecke algebras. These representations lead eventually to the definition of the Jones polynomial and the Homflypt polynomial of knots. Let $A = \mathbb{Z}[q^{\pm 1}]$ denote the ring of Laurent polynomials in q.

The **Hecke algebra**, H_n, is the algebra over A, with unit, generated by $1, T_1, \ldots T_{n-1}$ and with the following relations:

(1) $T_i T_j = T_j T_i$ for $|i - j| \geq 2$;
(2) $T_i T_j T_i = T_j T_i T_j$ for $|i - j| = 1$;
(3) $T_i^2 = (q - 1)T_i + q$ for all $1 \leq i \leq n - 1$.

The first two relations are the same relations as in the classic braid group. The last relation assures that the map given by $T_i \mapsto q$ is an A-algebra homomorphism. Moreover, it implies that each T_i has an inverse given by $T_i^{-1} = q^{-1}(1 - q + T_i)$.

Exercise A.1. *Show that the map given by $T_i \mapsto q$ is an A-algebra homomorphism from H_n to A.*

As before, we get natural inclusions $A = H_1 \subset H_2 \subset \cdots \subset H$, where H denotes the direct limit of the H_n.

As an A-module, H is generated by the monomials in the T_i. In fact, we can prove a sharper result: as an A-module, H_n is generated by a set of monomials in $T_1, \ldots, T_{n-1}$ with at most one occurrence of T_{n-1}. Firstly, consider the following sets Σ_i defined inductively by

$$\Sigma_i = \{1\} \cup T_i \Sigma_{i-1}, \ \Sigma_1 = \{1, T_1\}$$

so

$$\Sigma_1 = \{1, \ T_1\}$$
$$\Sigma_2 = \{1, \ T_2, \ T_2 T_1\}$$
$$\cdots$$
$$\Sigma_{n-1} = \{1, \ T_{n-1}, \ T_{n-1} T_{n-2}, \ldots, \ T_{n-1} \ldots T_1\}$$

A monomial m in the form $m = m_1 \ldots m_{n-1}$ with $m_i \in \Sigma_i$ is said to be in **normal form**.

Lemma A.1. *Monomials in normal form generate H_n.*

This lemma may be proved by induction on n using the relations of H_n.

As an easy consequence of this lemma, we derive the following.

Corollary A.1. $\dim_A(H_n) \leq n!$.

In fact, we will show that $\dim_A(H_n)$ is actually equal to $n!$. In order to achieve this goal, we have to make some preparations.

Let S_n denote the symmetric group on n symbols. The transposition of the ith and $(i+1)$th symbol is written as s_i or $(i, i+1)$ in cyclic notation. Denote the free A-module generated by the elements of the symmetric group S_n by AS_n. We obviously have

$\dim_A AS_n = |S_n| = n!$. We define a left operator $L_i \in \text{End}(AS_n)$ by setting

$$L_i(\pi) = \begin{cases} s_i\pi & \text{if } \ell(s_i\pi) > \ell(\pi) \\ (q-1)\pi + qs_i\pi & \text{otherwise} \end{cases}$$

for any $\pi \in S_n$ and then by linear extension. Here, ℓ is the reduced word length in the generators s_i.

It is easy to check that $L_i^2 = (q-1)L_i + q$.

In the same way, we define a right operator $R_j \in \text{End}(AS_n)$ by setting

$$R_j(\pi) = \begin{cases} \pi s_j & \text{if } \ell(\pi s_j) > \ell(\pi) \\ (q-1)\pi + q\pi s_j & \text{otherwise} \end{cases}$$

As in the case of L_i we easily calculate that $R_j^2 = (q-1)R_j + q$. Moreover, we get the following lemma, which shows that the left and right actions commute.

Lemma A.2. *For any $1 \leq i, j \leq n-1$, we have $L_i R_j = R_j L_i$.*

Proof. This lemma follows from a case-by-case verifications. In fact, six cases have to be considered. These are as follows:

Case 1 $\ell(s_i\pi s_j) = \ell(\pi) + 2$
Case 2 $\ell(s_i\pi s_j) = \ell(\pi) - 2$
Case 3 $\ell(s_i\pi s_j) = \ell(\pi)$ $\ell(s_i\pi) > \ell(\pi)$ $\ell(\pi s_j) < \ell(\pi)$
Case 4 $\ell(s_i\pi s_j) = \ell(\pi)$ $\ell(s_i\pi) < \ell(\pi)$ $\ell(\pi s_j) > \ell(\pi)$
Case 5 $\ell(s_i\pi s_j) = \ell(\pi)$ $\ell(s_i\pi) > \ell(\pi)$ $\ell(\pi s_j) > \ell(\pi)$
Case 6 $\ell(s_i\pi s_j) = \ell(\pi)$ $\ell(s_i\pi) < \ell(\pi)$ $\ell(\pi s_j) < \ell(\pi)$

The proof of Cases 1–4 is straightforward. For example, in Case 1, it is clear that $\ell(s_i\pi) > \ell(\pi)$ and $\ell(\pi s_j) > \ell(\pi)$ so $L_i R_j(\pi) = s_i R_j(\pi) = s_i\pi s_j = R_j L_i(\pi)$ because of associativity.

Cases 5 and 6 are trickier to prove and we provide the proof of 5 using Fig. A.1, the proof of 6 being similar.

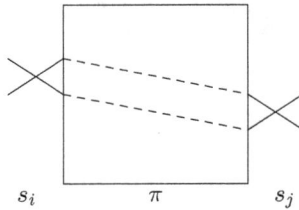

Fig. A.1. $\sigma_i \pi \sigma_j = \pi$.

(Case 5) Since $\ell(s_i \pi s_j) = \ell(\pi)$ and $\ell(s_i \pi) > \ell(\pi)$, $\ell(\pi s_j) > \ell(\pi)$, it follows that as products of the generators, the word $s_i \pi s_j$ is unreduced, but the words $s_i \pi$ and πs_j are reduced if π is written as a reduced word.

It follows that $s_i \pi s_j$ can be pictured as in the diagram with strings crossing by s_i and s_j but not crossing inside π. These crossings cancel, and so, $\sigma_i \pi \sigma_j = \pi$ and the case follows. Commutativity follows as before by associativity of functions. □

In the following lemma, the relationship between the operation L_i and the Hecke algebra H_n becomes clear.

We have the following.

Lemma A.3.

(1) $L_i L_j = L_j L_i$ for $|i - j| > 1$;
(2) $L_i L_j L_i = L_j L_i L_j$ for $|i - j| = 1$;
(3) $L_i^2 = (q - 1)L_i + q$ for all $1 \le i \le n - 1$.

Proof. Let π be an element of the symmetric group S_n and let π have length k. Let $\pi = s_{i_1} \cdots s_{i_k}$ be a reduced form of π. This implies that $\pi = R_{i_k} \cdots R_{i_1}(1)$. Set $R = R_{i_k} \cdots R_{i_1}$. Now, we can compute 1 for $|i - j| > 1$:

$$L_i L_j(\pi) = L_i L_j R(1) = R L_i L_j(1) = R(s_i s_j)$$

$$= R(s_j s_i) = R L_j L_i(1) = L_j L_i R(1) = L_j L_i(\pi)$$

For 2, a similar computation shows that $L_i L_j L_i = L_j L_i L_j$ for $|i - j| = 1$. The last equality of the lemma has already been shown. □

The last lemma shows that there is a well-defined algebra homo-morphism, $\rho : H_n \to End(AS_n)$ with $\rho(T_i) = L_i$.

Now, let $M = \{m_1, \ldots, m_{n!}\}$ be the set of monomials in H_n in normal form. We easily see that the set $\{\rho(m_1)(1), \ldots, \rho(m_{n!})(1)\}$ is linearly independent in AS_n. Hence, the set of monomials, M, is linearly independent. Thus, $dim_A(H_n) \geq n!$. Together with Corollary A.1, we obtain the following.

Corollary A.2. $dim_A(H_n) = n!$.

This corollary implies that the monomials in normal form actually form a basis of H_n. We can conclude that any element $c \in H_{n+1}$ can be written in the form $c = a + \sum_i a_i T_n b_i$ with $a, a_i, b_i \in H_n$. Moreover, this representation is unique up to equivalence $xyT_n z = xT_n yz$ if $y \in H_{n-1}$. Precisely, we have the following.

Theorem A.1. *The algebra H_{n+1} can be written as the following combination:*

$$H_{n+1} \cong H_n \oplus H_n \otimes_{H_{n-1}} H_n.$$

A.2.1 Trace functions

We will use the results of the last section in order to define trace functions on H_n. With the aid of these trace functions, we will be able to define the Homfly Polynomial of a link.

Let R be an A-algebra with unit and let S be an A-module with unit. A **trace function** $\tau : R \to S$ is defined to be an A-linear map which satisfies $\tau(ab) = \tau(ba)$. It is called **normalized** if $\tau(1) = 1$.

Theorem A.2. *There is a unique normalized trace function $\tau :$ $H \to A[z]$ such that for all $a, b \in H_n : \tau(aT_n b) = z\tau(ab)$.*

Proof. Since τ is A-linear, τ has to be the identity on $H_0 \cong A$. The formula

$$\tau(aT_n b) = z\tau(ab)a, b \in H_n$$

gives us a unique way in which we can extend τ from H_n to H_{n+1}. So, we only have to show that $\tau(ab) = \tau(ba)$.

By the structure theorem, we need only show that $\tau(m_1m_2) = \tau(m_2m_1)$ where m_1 and m_2 are monomials in normal form.

Suppose m_1 but not m_2 contains T_n. So, $m_1 = m_1'T_nm_1''$, where m_1', m_1'' and m_2 lie in H_n. Then

$$\tau(m_1m_2) = \tau(m_1'T_nm_1''m_2) = z\tau(m_1'm_1''m_2) \text{ by definition}$$

$$= z\tau(m_2m_1'm_1'') \text{ by induction}$$

$$= \tau(m_2m_1'T_nm_1'') = \tau(m_2m_1)$$

Now, suppose that both m_1 and m_2 contain T_n. Let

$$m_1 = m_1'T_nm_1'' \quad \text{and} \quad m_2 = m_2'T_nm_2'' \quad \text{in normal form}$$

We will use the two general identities

$$\tau(\mu_1T_n\mu_2T_n) = \tau(T_n\mu_1T_n\mu_2)$$

and

$$\tau(\mu_1T_n\mu_2T_n\mu_3) = \tau(\mu_3\mu_1T_n\mu_2T_n)$$

where μ_1, μ_2, μ_3 are monomials in H_n.

If we assume that these are true, then we have

$$\tau(m_1m_2) = \tau(m_1'T_nm_1''m_2'T_nm_2'')$$

$$= \tau(m_2''m_1'T_nm_1''m_2'T_n) \quad \text{identity 2}$$

$$= \tau(T_nm_2''m_1'T_nm_1''m_2') \quad \text{identity 1}$$

$$= \tau(m_2'T_nm_2''m_1'T_nm_1'') \quad \text{identity 2}$$

$$= \tau(m_2m_1)$$

To prove identity 1, we may assume that μ_1 and μ_2 are in normal form, say, $\mu_1 = \mu_1'T_{n-1}\mu_1''$ and $\mu_2 = \mu_2'T_{n-1}\mu_2''$. Then

$$\tau(\mu_1T_n\mu_2T_n) = \tau(\mu_1T_n\mu_2'T_{n-1}\mu_2''T_n)$$

$$= \tau(\mu_1\mu_2'T_nT_{n-1}T_n\mu_2'')$$

$$= \tau(\mu_1\mu_2'T_{n-1}T_nT_{n-1}\mu_2'')$$

$$= z\tau(\mu_1\mu_2'T_{n-1}^2\mu_2'')$$
$$= z(q-1)\tau(\mu_1\mu_2'T_{n-1}\mu_2'') + zq\tau(\mu_1\mu_2'\mu_2'')$$
$$= z(q-1)\tau(\mu_1\mu_2) + zq\tau(\mu_1'T_{n-1}\mu_1''\mu_2'\mu_2'')$$
$$= z(q-1)\tau(\mu_1\mu_2) + zq\tau(\mu_1'\mu_1''\mu_2'T_{n-1}\mu_2'')$$
$$= \tau(T_n\mu_1T_n\mu_2)$$

Identity 2 is proved similarly.

The trace function is clearly normalised and so the inductive step is proved. □

A.2.2 *The classic braid group and its relation to the Hecke algebra*

The various generalised braid groups were considered in Chapter 6. The classic braid group, B_n, is generated by $\{\sigma_1, \ldots, \sigma_{n-1}\}$ and has the following relations:

(1) $\sigma_i\sigma_j = \sigma_j\sigma_i$ for $|i-j| > 1$;
(2) $\sigma_i\sigma_j\sigma_i = \sigma_j\sigma_i\sigma_j$ for $|i-j| = 1$.

These are the same as the first two relations of the Hecke algebra with σ replacing T. In other words, H_n is obtained from the group ring $A[B_n]$ by adding the relation

$$\sigma_i^2 = (q-1)\sigma_i + q \quad \text{for all } 1 \le i \le n-1$$

This can be rewritten as

$$\sigma_i - q\sigma_i^{-1} + 1 - q = 0$$

Let $\beta^+ = u\sigma_i v$ be a braid, $\beta^- = u\sigma_i^{-1}v$ and $\beta^0 = uv$, where $u, v \in B_n$. If $(\hat{\ })$ denotes knot closure, then

$$K^+ = \hat{\beta^+}, \quad K^- = \hat{\beta^-}, \quad K^0 = \hat{\beta^0}$$

is a skein triple.

A.2.3 *Hecke algebras and braids*

We have seen that the Hecke algebra is obtained from the group ring $A[B_n]$ by adding the quadratic relation. So, there is a canonical representation $\phi : B_n \to H_n$ given by $\phi(\sigma_i) = T_i$. Taking the composition of ϕ and τ, we obtain a trace function $t : B_n \to A[z]$. We want to use this trace function to define an invariant of oriented links. To do this, we briefly recall the theorems of Alexander and Markov given in Chapters 7 and 8.

The theorem of Alexander asserts that any oriented link may be obtained as the closure of a braid, i.e. given a knot K, there is a braid b such that the closure of b is isotopic to K.[20]

Thus, Alexander's theorem concerns the surjectivity of the closing operation. Markov's theorem deals with the injectivity of this operation.

Theorem A.3 (Markov's Theorem). *Two classical braids b and b' become the same link after closing if and only if they can be related by a finite sequence of Markov moves. Assume $b, g \in B_n$. These moves are as follows:*

Conjugation $b \leftrightarrow g^{-1}bg$;
Stabilisation $b \leftrightarrow b\sigma_n^{\pm 1}$.[21]

If we can find an invariant of braids which is also invariant under the above moves, then we will have a knot-invariant.

A.2.4 *Defining the Homflypt polynomial*

We want to make the trace t, defined on braids, into a link invariant. By the property of trace functions, t is already invariant under conjugation. It is not yet invariant under the stabilisation, but with the help of a normalization factor, this can be achieved. Define $e : B_n \to \mathbb{Z}$ by $e(\sigma_i) = 1$ for all i. It is easy to see that e is a well-defined group homomorphism. (In fact, it is the abelianization of B_n and will also define the writhe of the knot closure.)

---

[20] Assuming the knot theory is regular.
[21] In Chapter 8, we showed that the exchange move was unnecessary.

Theorem A.4. *The Homfly polynomial is the link-invariant V defined by*

$$V(K) = z^{-(n+e(b)-1)/2} \left(\frac{q}{1+z-q} \right)^{(n-e(b)-1)/2} t(b),$$

where b is a braid whose closure is isotopic to K.

Proof. In the literature, the Homflypt polynomial is given with several different parametrizations and solutions to the skein relation. We will use the skein relation defined in Chapter 9. If we can show that the definition above satisfies this skein relation for suitable values of q and z, then the proof will follow.

Put

$$\lambda = \left(\frac{-z}{z+1-q} \right)^{1/2}, \quad \mu = i(q^{-1/2} - q^{1/2})$$

Let L_+, L_-, and L_0 be links, such that their link diagrams only differ in a small disc where they look as shown in Fig. 9. Then we have

$$\lambda V(L_+) + \lambda^{-1} V(L_-) + \mu V(L_0) = 0.$$

As indicated before, Markov's theorem assures that the Homfly polynomial is well defined. □

Appendix B

Categories

B.1 Introduction

In this appendix, give a brief introduction to category theory and in particular to groupoids. This is because one way to look at a knot theory is as a groupoid.

B.2 Defining the concepts

A **category** is an algebraic structure comprising **objects** that are linked by **morphisms** or **arrows**. A category has two basic properties: the ability to compose the morphisms with associativity and the existence of an identity morphism for each object.

More specifically, if $f : A \to B$ and $g : B \to C$ are morphisms between objects A to B and objects B to C, then f and g can be composed to form the morphism $g \circ f$ from A to C. Associativity means that

$$h \circ (g \circ f) = (h \circ g) \circ f$$

when composition is possible.

Moreover, every object A has an **identity** morphism $1_A : A \to A$ such that

$$1_B \circ f = f = f \circ 1_A$$

A simple example is the **category of sets**, whose objects are sets and whose morphisms are functions between sets.

Exercise B.1. *In a (male) army camp, the barber shaves everyone who cannot shave themselves. Who shaves the barber? What is this to do with category theory? Why is the category of sets not a set?*

Category theory is a branch of mathematics that seeks to generalize much of mathematics in terms of categories, independent of what their objects and morphisms represent. Many branches of modern mathematics can be described in terms of categories and in doing so often reveal deep insights and similarities between seemingly different areas of mathematics. As such, category theory provides an alternative foundation for mathematics to set theory and other proposed axiomatic foundations. In general, the objects and morphisms may be abstract entities of any kind, and the notion of category provides a fundamental and abstract way to describe mathematical entities and their relationships.

In addition to formalizing mathematics, category theory is also used to formalize many other systems in computer science, such as the semantics of programming languages.

Exercise B.2. *Find examples of categories in group theory, linear algebra and elsewhere.*

A morphism $f : A \to B$ is called an **isomorphism** if there is a morphism $g : B \to A$ such that $g \circ f = 1_A$ and $f \circ g = 1_B$. We write f^{-1} for g if it exists. If $A = B$, then an isomorphism is called an **automorphism**. A category in which all morphisms are isomorphisms is called a **groupoid**.

Exercise B.3. *Show that a group G is a groupoid with one object, G and a morphism for each element $g \in G$. Composition for morphisms is the same as composition of the group elements.*

Let $\mathcal{G}$ be a groupoid and A an object of $\mathcal{G}$. Then the **point group** of A is the set of morphisms which start and finish at A, i.e. are automorphisms. Composition in this group is the same as morphism composition.

Example B.1. Let X be a topological space. The **fundamental groupoid**, $\Pi(X)$, of X is the category with object, the points of X and morphisms homotopy classes of paths fixed at their endpoints. The composition is the usual composition of paths. The point group at $x_0 \in X$ is the **fundamental group** of X, $\pi_1(X, x_0)$ based at x_0.

Let $\mathcal{C}$ denote a category and let $\mathcal{R}$ be a subcollection of the morphisms of $\mathcal{C}$. We say that $\mathcal{R}$ **generates** $\mathcal{C}$ if given a morphism $f : A \to B$ of $\mathcal{C}$, there is a finite set $r_i : A_i \to A_{i+1}$, $i = 0, \ldots, n$ of members of $\mathcal{R}$ such that

$$A_0 = A, \quad A_{n+1} = B, \quad \text{and} \quad f = r_n \circ \cdots r_0$$

Exercise B.4. *Let V be a vector space with scalar field, the complex numbers, say. Then V can be considered as a groupoid with one object and morphisms the linear automorphisms of V. Show that the elementary automorphisms generate V, where the elementary automorphisms are defined by two types of matrices:*

$$E_1 = \begin{pmatrix} 1 & 0 & \cdots & & 0 \\ 0 & \ddots & \cdots & & 0 \\ 0 & \cdots & \lambda & \cdots & 0 \\ 0 & \cdots & & \ddots & 0 \\ 0 & 0 & \cdots & & 1 \end{pmatrix} \quad \text{and}$$

$$E_2 = \begin{pmatrix} 1 & 0 & \cdots & & 0 \\ 0 & \ddots & \cdots & & 0 \\ 0 & \cdots & 1 & \mu & 0 \\ 0 & \cdots & & \ddots & 0 \\ 0 & 0 & \cdots & & 1 \end{pmatrix}$$

The matrix E_1 is the identity matrix with one diagonal entry replaced by a non-zero number λ. The matrix E_2 is the identity matrix with an off-diagonal entry replaced by μ.

We are now ready to make a fancy definition of a (generalized) **knot theory**. This is the groupoid in which the objects are diagrams and the morphisms are generated by the R moves. A (generalized) **knot** is a path component of the groupoid.

Appendix C

Coding Knots on Surfaces

C.1 Diagram codes

By a *code* in this situation, we mean any finite bits of information which define the curve. In this chapter, we will associate to certain curves a **code** consisting of a signed permutation in cycle form. This idea goes back to Gauss and has since been modified by Dowker and Thistlethwaite and Bartholomew [44], however we will give our own slant on the procedure. In particular, we will number the edges of the curve in order, which leads to a numbering of the crossings. Other authors work in the opposite direction and number the crossings first.

It seems a remarkable fact that a 3-dimensional object such as a classical knot can be reduced to a 2-dimensional object like a diagram and then this can be further reduced to a 1-dimensional piece of code. However, on mature reflection, this is not so remarkable, as we are not interested in a rigid wire model of a knot, which would require an infinite amount of information to describe it, but something more like a knot made of rope which can be flexibly manipulated and manoeuvred in space. We have already seen that a knot can be described by a chord diagram which is essentially 1-dimensional.

C.2 Signed permutations and cycles

In this section, we look at signed permutations and their properties. This is because of the relationship between signed permutations and knot diagrams. Do not despair if you cannot see where all this investigation is leading, all will be revealed soon.

A **signed permutation** is best represented as a square matrix with only one non-zero entry, ± 1, in each row or column. As with normal permutations, each element has a representation as a product of disjoint **signed permutation cycles**. Such a cycle being a sequence

$$(x_1 y_1 x_2 y_2 \ldots, x_n y_n)$$

where x_i is an integer in the range $1, 2, \ldots, n$ for some n and y_i is either a plus or a minus.

All the integers in the range are used because this is a permutation, but with two types of moves taking one integer to another. We can think of the moves as a badly shuffled pack of n cards with a plus if the card is face down and a minus if the card is face up.

A cycle of length n is a made up of n triples $x_k y_k x_{k+1}$, $k = 1, \ldots, n \bmod n$, of the form $i_{+}j$ or $i_{-}j$. These are shorthand for the moves of the permutation. For example, $i_{+}j$ means that the integer i goes to the integer j in a positive manner and $i_{-}j$ means that the integer i goes to the integer j in a negative manner, whatever that means. In examples, we will sometimes replace $i_{+}j$ with ij for reasons of space.

For example, $(1_{+}2_{+}3_{-})$ or equivalently (123_{-}) means 1 is sent to 2 which is sent to 3 all positively, while 3 is sent back to 1 negatively.

Disjoint cycles of total length n and an $n \times n$ signed permutation matrix, M, are related by the following formula. We have a number of moves, $i_{+}j$ or $i_{-}j$. So, for $i, j \in \{1, 2, \ldots n\}$,

$$M_{ij} = \begin{cases} 1 & \text{if } i_{+}j \\ -1 & \text{if } i_{-}j \\ 0 & \text{else} \end{cases}$$

So, an $n \times n$ signed permutation matrix is exactly equivalent to a signed permutation of the n numbers $\{1, 2, \ldots n\}$.

The diagonal entries represent fixed points and come in two flavours, say, $(i_+) = (i)$ or (i_-) and represent a diagonal entry of the matrix in the ith position as $+1$ or -1, respectively.

Exercise C.1. *Show that the signed permutation* $(1_+3_-2_-)(4_-)$ *is related to the* 4×4 *signed permutation matrix,*

$$\begin{pmatrix} 0 & 0 & 1 & 0 \\ -1 & 0 & 0 & 0 \\ 0 & -1 & 0 & 0 \\ 0 & 0 & 0 & -1 \end{pmatrix}$$

Under matrix multiplication, the set of all $n \times n$ signed permutation matrices form a group written $\tilde{S}_n$. The subset S_n of all positive permutations can be identified with the symmetric group of n objects.

There is a natural retraction map $\tilde{S}_n \rightarrow S_n$ given by making the negative entries positive. Hence, there is a split short exact sequence,

$$1 \rightarrow C_2^n \rightarrow \tilde{S}_n \rightarrow S_n \rightarrow 1$$

where C_2^n is represented by $n \times n$ diagonal matrices with ± 1 entries.

Exercise C.2. *Write down all eight elements of* $\tilde{S}_2$ *and its multiplication table.*

Exercise C.3. *Show that* $\tilde{S}_2$ *is isomorphic to* D_4, *the dihedral group of symmetries of a square. Hint:* $(1)(2_-)$ *is a reflection and* (12_-) *is a rotation through* $\pi/2$.

Exercise C.4. *Let* $x = (123_-)$. *Show that* $x^3 = -1$. *Deduce that* x *has order 6.*

Signed permutations can also be defined as *quipu*, see the work of Bae *et al.* [19]. In Fig. C.1, the signed permutation $(13_-)(2_-4)$ is represented as a quipu.

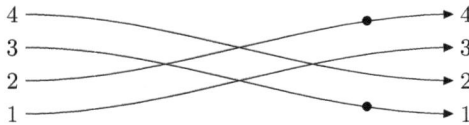

Fig. C.1. $(13_-)(2_-4)$.

Readers who have read Chapter 6 will have no trouble under-standing Fig. C.1. The arc from 2 to 4 and the arc from 3 to 1 have a sliding bead which indicates a negative move. Multiplication is concatenation as usual and if two beads meet they self-destruct.

Exercise C.5. *Use the quipu method to show that the permutation* $(13_-)(2_-4)$ *is a square root of* -1.

C.3 Shadow diagrams and signed permutations

Recall that a shadow diagram is the immersed image of several circles (components) in the plane or sphere in general position. So, it is a pre-diagram of a knot before information is added at the vertices to make it into a knot diagram. However, for simplicity, we shall just call a shadow diagram a diagram and denote it typically by D. We shall show that there is an intimate relationship between a diagram D and a finite number of signed permutations.

Firstly, assume that D has just one component. An **edge numbering or labelling** of D, denoted $\mathcal{L}$, is defined by picking an edge, labelling it 1 and continue 2, 3, etc. for the consecutive edges defined by the orientation. The numbering is defined mod $2n$, where n is the number of vertices.

Consider the situation of the numberings at the two vertices shown in the Fig. C.2, where the incoming and outcoming edges have even and odd labels.

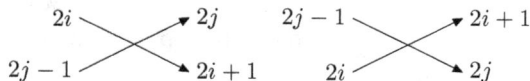

Fig. C.2.   Crossings of types I and II.

Crossings like this are divided into type I where the even labelled edges are above the odd labelled edges and type II which is the reverse.

Theorem C.1. *All crossings are either of type I or II.*

Proof. Note that this proof depends on the diagram being in the plane or sphere. If the diagram lies on another surface, the theorem need not be true. In order to prove this theorem, we need the following concept. □

C.3.1 *Chessboard colouring*[22]

A **chessboard colouring** of the regions of a diagram is a colouring by white and black so that at each crossing the colours alternate about the vertex. This means that the regions in the neighbourhood of edges are coloured black or white so that opposite colours are on either side of an edge border.

Lemma C.1. *Any diagram can be chessboard coloured.*

Proof. Let P_0 be a point in the unbounded region of a diagram, D. Let P be a point in one of the regions of D. Join P to P_0 by a path which avoids the crossing points of D and crosses the edges of D transversely at the points $P_1, \ldots, P_k$, say. If k is odd, call P an **odd** point. Otherwise, it is an **even** point.

It is not hard to see that the following two facts are true: (1) the parity of the point P is independent of the path chosen and (2) all points in the same region have the same parity. The first fact is because any two paths joining P to P_0 are homotopic keeping P and P_0 fixed, since the diagram is in the plane and that this homotopy in general position will not change the parity. The second fact is that any two points in a region can be joined by a path disjoint from the diagram and the composition of this path with a path to P_0 crosses the edges of D at the same points and so have the same parity.

---

[22]Some authors call this checkerboard colouring.

This means we can call a region odd or even according to the parities of its points. We colour the odd regions black and the even regions white. □

Note that this statement is not necessarily true for curves on surfaces other than the plane or sphere. For example, consider the **Hopf curve** consisting of the meridian and longitude of a torus. Nevertheless, even this can be represented by the code using the power of virtual crossings.

Exercise C.6. *Show that one can chessboard colour a diagram on a surface S if and only if it represents zero in the first mod 2 homology group, $H_1(S; \mathbb{Z}_2)$.*

Now, we return to the proof of Theorem C1. Note that there are two possible chessboard colourings and we choose the one in which black is to the left of edge numbered 1. A moment's reflection will convince you that this is true for all odd numbered edges and that even numbered edges have black on the right.

This means that the incoming edges either span a black or white wedge and this defines the two types, I and II (see Fig. C.3).

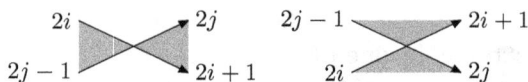

Fig. C.3. i_{+j} and i_{-j} associated with types I and II.

C.4 Making the code

We have seen using chessboard colouring and a consecutive labelling of the edges that every crossing is of type I or II and has integers $2i$ and $2j - 1$ associated with the incoming edges. The integers $2i$ and $2j - 1$ are taken mod $2n$ and the integers i and j are taken mod n. We call the incoming edge labelled $2i$ the **naming edge** as it could be used to number the crossing with V_i. The one incoming edge with odd labelling $2j - 1$ is called the **permutation edge**.

We therefore have a set of edges E_k where k is an integer mod $2n$ and a set of crossing points or vertices V_i using the naming edge E_{2i}, where i is an integer mod n.

Associated to the crossing V_i is the signed permutation π from the permutation edge E_{2j-1}, where j is an integer mod n. The sign of the permutation is taken according to the type of the crossing as follows:

$$i_{+}j, \quad \text{if type I, } i_{-}j, \quad \text{if type II}$$

Let us illustrate this with the example in Fig. C.4. We colour the unbounded region, the internal bigon and the internal trigon black.

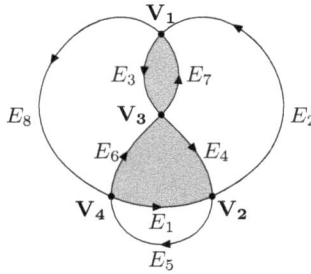

Fig. C.4. The figure 8 curve.

The curve is a figure 8 with four crossings, $V_1, \ldots, V_4$, of which two are of type I and two are of type II. Consider the crossing V_1. The incoming edges are numbered 2 and 7 and is of type II, so we get the move $1_{-}4$. The remaining moves are $2_{+}1$, $3_{-}2$ and $4_{+}3$.

The matrix is

$$\begin{pmatrix} 0 & 0 & 0 & -1 \\ 1 & 0 & 0 & 0 \\ 0 & -1 & 0 & 0 \\ 0 & 0 & 1 & 0 \end{pmatrix}$$

corresponding to the signed permutation $(1_{-}43_{-}2)$.

C.5 Changing the starting point and the push map

The initial edge labelled 1 was chosen at random. If we call this labelling $\mathcal{L}_1$, then we have a sequence of possible labellings

$$\mathcal{L}_1, \mathcal{L}_2, \ldots, \mathcal{L}_{2n-1}$$

where $\mathcal{L}_k$ is obtained from $\mathcal{L}_{k-1}$ by labelling the edge E_2 as E_1 and so on consecutively, $k = 1, 2, \ldots$, mod $2n$.

This process, $\mathcal{L}_k \rightarrow \mathcal{L}_{k+1}$ is called a **push**. The push changes the signed permutation, π_k to π_{k+1}.

If instead we start at the edge previously labelled by 2, then the black regions become white and vice versa. So, if before there were p crossings of type I and q crossings of type II, afterwards there would be q crossings of type I and p crossings of type II. However, the numbering of the crossings and the moves will change.

If we apply the push to our example, then everything changes. Note that the infinite region is now white (see Fig. C.5).

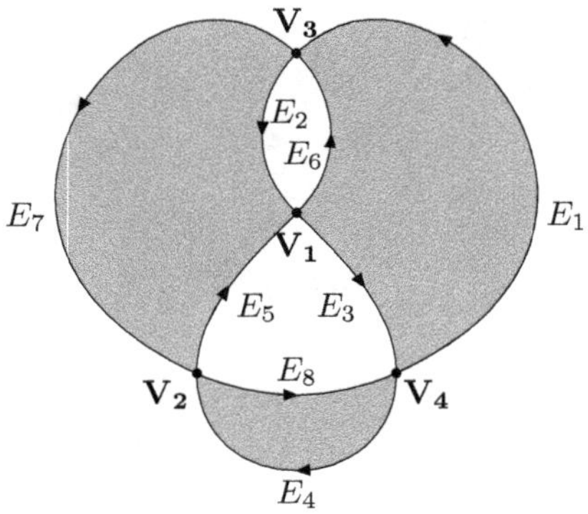

Fig. C.5.   The figure 8 curve, again.

The code is now the product of two 2-cycles, $(1_+3_+)(2_-4_-)$.

If we push one further and start counting at the edge originally labelled 3, then the colouring reverts to the original and the signed permutation becomes $(1_+4_-3_+2_-)$. This is the same as the original permutation with the signs reversed.

The general change is easily summed up by the following lemma.

Lemma C.2. *Changing the initial edge to the next edge changes the vertex V_i to $V_j = V_{\pi(i)}$ and swaps its type. In particular, i_+j becomes $j_-i + 1$ and i_-j becomes $j_+i + 1$.*

This shows that the code depends not only on the plane figure but also on the starting point and how the regions are shaded.

Exercise C.7. *Using Lemma C.2., find all possible eight codes for the figure 8 curve.*

Exercise C.8. *Show that for the trefoil curve the code is either $(1_+3_+2_+)$ or $(1_-3_-2_-)$ depending on the shading.*

Exercise C.9. *Show that there are $2^n(n-1)!$ possible codes for a curve with one component and n crossings.*

For curves with more than one component, the code has to be modified as follows. Start with one component curve, c_1, say, and start labelling the edges in order from an arbitrary edge of c_1 with a black coloured region on its left. This time some of these crossings will be self-crossings of c_1 and others will be where different components cross c_1. When all the edges of c_1 are numbered $E_1 \ldots, E_{2n}$ in order, start on another component, c_2 say, and choose and label an edge E_{2n+1} which has a black coloured region on its left. Continue labelling c_2 and so on in this fashion until all the edges of the diagram are labelled, $E_1, \ldots, E_{2N}$, for some integer N. As in the single component case, we get a signed permutation. But this time the numbering of the edges may be modulo some different integers.

Exercise C.10. *Consider Fig. C.6 where the edges of the Borromean rings are partially numbered. (1) Finish the numbering. (2) Colour in the black regions. (3) Show that there are an equal number of type I and type II crossings. (4) Find the signed permutation.*

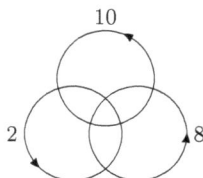

Fig. C.6. Edges of the Borromean rings partially numbered.

C.6 The reverse procedure

We now show how the information supplied by a signed permutation, may be used to define a curve on an orientable surface. The procedure is the reverse of the previous discussion. Moreover, the genus of the surface may easily be calculated, and in particular, we will know which ones are planar. To simplify the exposition, we will assume that the permutation generates a diagram which has only one component. We leave the more general case as an exercise for the reader.

We assume we have a signed permutation of degree n, so it will come as no surprise to the reader that we will generate an oriented curve with n crossings and $2n$ edges which has, at least locally, a chessboard shading. The crossings will be contained within n squares. The edges of the squares are then joined by $2n$ ribbons according to the information contained in the permutation.

The two types of squares, Q_{ij} and Q'_{ij}, are illustrated in Fig. C.7. The square Q_{ij} has two input edges numbered $2i, 2j - 1$ and two output edges numbered $2j, 2i + 1$. The square Q'_{ij} has the same but in the opposite order.

Fig. C.7.   Spare parts, squares.

The ribbons, which are oblongs, have two short edges available to stick to the squares and are half-coloured black (see Fig. C.8). They can be flipped over the long oriented line of symmetry so that the black half is either below or above in the diagram. Remarkably, there is no need for a half-twist in the patching procedure.

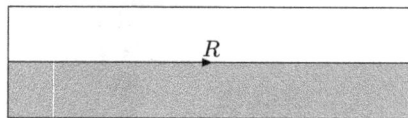

Fig. C.8.   Spare parts, ribbons.

Suppose we have the information in the form of an $n \times n$ signed permutation and hence a number of moves of the form $i_{+}j$ or $i_{-}j$. To each move of the form $i_{+}j$, we pull down from the component shelf the square Q_{ij} and to each pair of the form $i_{-}j$, we pull down a Q'_{ij}.

We then gather $2n$ ribbons and consult the handbook. Since the information comes from a signed permutation, for each Q_{ij}, there is one of $Q_{k,i+1}, Q'_{k,i+1}$ and one of Q_{jl}, Q'_{jl} for some k, l which can be joined by ribbons so that the orientation is continuous and the coloured sub-squares match. It follows that we have an oriented surface which is the regular neighbourhood of a graph with n vertices and $2n$ edges.

It is a fact that a connected compact orientable surface of genus g and with c boundary circles has Euler–Poincaré characteristic $2 - 2g - c$, see the work of Massey [219].

Our graph has Euler–Poincaré characteristic, the difference between the number of vertices and edges, i.e. $-n$, which will be the same as our constructed surface. When we know c, we will know $g = (2 + n - c)/2$ which will be the genus of our surface when the boundary circles are filled in with c disks.

To calculate c, we construct two (unsigned) permutations, π_W and π_B, of the integers $\{1, 2, \ldots, 2n\}$. The two permutations are labelled **white** and **black** and the motivation for their definitions is as follows. Each ribbon has a white long edge and a black long edge. The numbering of the ribbon and their edges is defined by the numbering of the internal edge. If the ribbon belongs to an even-numbered internal edge, then orient the black edge in the same direction as the internal edge and orient the white edge in the opposite direction. For an odd numbered internal edge, do the reverse. The oriented edges all join up to make, say, c_W white circles and c_B black circles.[23]

The total number of boundary circles is

$$c = c_W + c_B$$

which are the number of cycles of the permutations π_W and π_B. These permutations are given by the following rule based on the squares:
If Q_{ij}, then

$$
\begin{array}{ccc}
2i & \leftarrow & 2j \\
\Downarrow & & \Uparrow \\
2j - 1 & \rightarrow & 2i + 1
\end{array}
$$

---

[23] In the literature, these are sometimes called right/left turning circles.

If Q'_{ij}, then

$$
\begin{array}{ccc}
2i & \Rightarrow & 2j \\
\uparrow & & \downarrow \\
2j-1 & \Leftarrow & 2i+1
\end{array}
$$

The single arrows represent the moves of the white permutation, π_W, and the double arrows represent the moves of the black permutation, π_B.

The justification for these formulæ can be found in Fig. C.9.

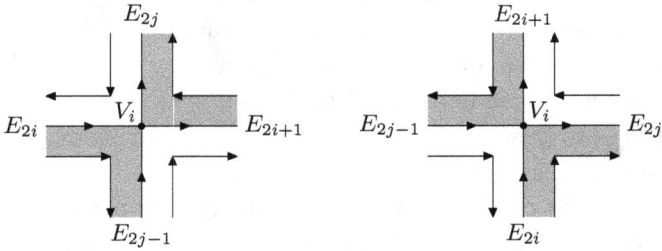

Fig. C.9.   Orienting the black and white boundary edges.

For example, consider the signed permutation $(1_-2_+3_-)(4_-)$. This is also represented by the 4×4 matrix:

$$
\begin{pmatrix}
0 & -1 & 0 & 0 \\
0 & 0 & 1 & 0 \\
-1 & 0 & 0 & 0 \\
0 & 0 & 0 & -1
\end{pmatrix}
$$

The moves and the working can be set out as follows:

$$
\begin{array}{cccc}
1_-2 & 2_+3 & 3_-1 & 4_-4 \\
2 \Rightarrow 4 & 4 \leftarrow 6 & 6 \Rightarrow 2 & 8 \Rightarrow 8 \\
\uparrow \quad \downarrow & \Downarrow \quad \Uparrow & \uparrow \quad \downarrow & \uparrow \quad \downarrow \\
3 \Leftarrow 3 & 5 \rightarrow 5 & 1 \Leftarrow 7 & 7 \Leftarrow 1
\end{array}
$$

So, the white permutation is $(1643278)(5)$ and the black permutation is $(17)(2456)(3)(8)$.

So, $c = 2 + 4 = 6$, $n = 4$ and the genus is $1 + (4 - 6)/2 = 0$. Hence, the curve is planar.

Exercise C.11. *Use the code cycle, $(1_+3_-2_-)$, to construct the (non-planar) graph in Fig. C.10. Use the above method to show that it lies on a surface of genus one, i.e. a torus.*

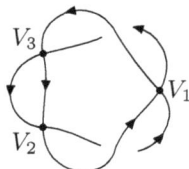

Fig. C.10. Graph corresponding to $(1_+3_-2_-)$.

C.6.1 *How to handle classic knots and beyond*

So far, the signed permutation codes a flat diagram. The graph with numbered edges in Fig. C.10 has the property that the pair of incoming edges are odd and even. However, to lift it into space or to code a general knot, we add tags to the details, keeping the information about the type of crossing. So, a **generalised signed permutation** is a product of cycles of the form

$$(x_1 y_1 x_2 y_2 \ldots, x_n y_n)$$

where x_i is an integer in the range $1, 2, \ldots, n$ for some n and y_i is either $+_a$ or $-_a$, where a is a tag and the sign indicates the type of crossing as usual.

So, a right-hand trefoil is coded as $(1_{+_r}3_{+_r}2_{+_r})$ and a left-hand trefoil is coded as $(1 +_{\bar{r}} 3 +_{\bar{r}} 2+_{\bar{r}})$.

Now, the Hopf curve mentioned at the beginning can be coded as a virtual knot. Consider Fig. C.11. Let V_1 be a virtual crossing and V_2 be a flat crossing. Both crossings are type I, so the generalised permutation cycle is

$$(1_{+_v}2_+)$$

and the generalised permutation matrix is

$$\begin{pmatrix} 0 & v \\ 1 & 0 \end{pmatrix}$$

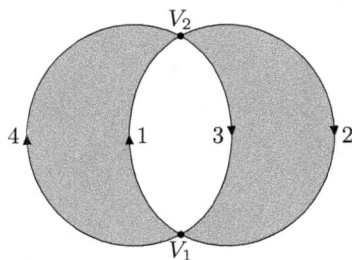

Fig. C.11.   The Hopf curve.

Examples of Planar Doodles

D.1 Introduction

In this appendix, we will give some examples of planar doodles, especially those which are part of an infinite sequence inspired by geometry. At the end of the appendix, there is a table of planar *prime* and *superprime* doodles up to 14 crossings investigated by the author and Bartholomew [10].

D.2 The Diophantine equation

This section looks at the restrictions which can be placed on planar doodles with regard to the size of their regions, etc.

Consider a doodle with n crossings. Each crossing contributes four half-edges. By counting them up, we see that e, the number of edges, satisfies $e = 2n$. If f is the number of faces and the doodle lies in a surface of genus g, then by the Euler–Poincaré equation

$$n - 2n + f = 2 - 2g$$

So, if the doodle is planar and the surface is a sphere, then $f = n + 2$. This count will include the infinite face containing the point at infinity.

We are only interested in minimal doodles, i.e. those which have no monogons or bigons. Suppose a minimal doodle diagram with

n crossings has f_i i-gon regions for $3 \leq i \leq p$. Then, by elementary counting of edges and region,

$$f_3 + f_4 + \cdots + f_p = n + 2 \qquad \text{(a)}$$

$$3f_3 + 4f_4 + \cdots + pf_p = 4n \qquad \text{(b)}$$

Conversely, a doodle diagram is partly characterised by a solution to the above for a given n. We call the combination of the number of crossings and a solution a *doodle code*.

For example, the code of the Borromean rings is

$$6 \text{ (six crossings)}, \quad 8 \text{ (eight trigons)}$$

Similarly, the code for the poppy is

$$8, 8, 2$$

since it has eight crossings, eight trigons and two 4-gons.

The following two lemmas place constraints on the numbers in the codes.

Lemma D.1. *The number of trigons of a minimal doodle with n crossings satisfies*

$$8 \leq f_3 \leq \frac{4n}{3}$$

and the number of i-gons where $i > 3$ satisfies

$$0 \leq f_i \leq \frac{4n - 24}{i}, \quad i > 3$$

Proof. If we multiply equation (a) above by 4 and subtract (b), we see that

$$f_3 = 8 + f_5 + 2f_6 + \cdots + (p - 4)f_p$$

where p is the size of the largest polygon region and the left inequality of the first condition follows. The right inequality of the first condition follows from (b). The proof of the second inequalities follows similarly. $\qquad\square$

Doodles have sort of addition similar to classical knots. If a doodle can be separated into two or more topological components by a simple closed curve, then a connecting bridge between these will reduce both the components and the topological components.

Doodles which cannot be formed in this manner are called **prime doodles**. Just because a doodle is unprime doesn't mean it is uninteresting. In Fig. D.1 is a non-trivial doodle which is the 'sum' of two trivial doodles. Something similar occurs in virtual knot theory with the Kishino knot, as shown by Fenn *et al.* [72].

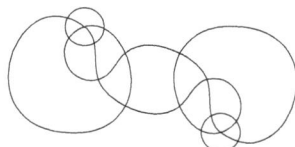

Fig. D.1. Non-trivial doodle.

Exercise D.1. *Show that if a minimal doodle is prime with n crossings, then every region has less than $(n+1)/2$ edges.*

Using Lemma 13.2.27 and the two equations (a) and (b), it is an easy matter to determine all possible values of f_i, the number of i-gons of doodles with a given number of crossings. The following two tables of possible codes up to $n = 11$ are easily found by hand. The final column entries are the corresponding number of doodles. Note that some codes do not give rise to doodles. For higher values of n, there are duplicates.

Number of crossings	Prime minimal planar doodles				
	f_3	f_4	f_5	f_6	No.
6	8				1
7	8	1			0
8	8	2			1
9	8	3			1
10	8	4			1
10	9	2	1		0
10	10	0	2		1
11	8	5			1
11	9	3	1		1
11	10	1	2		1
11	11	0	1	1	0

Exercise D.2. *Show that the only planar doodle with equal faces is the Borromean rings.*

D.3 Sequences of planar doodles

Here is a sequence of doodles $B_3, B_4, \ldots$ starting with the Borromean doodle, B_3 and the poppy, B_4. Let the vertices of the two concentric n-gons be $X_1 X_2 \ldots X_n$ and $Y_1 Y_2 \ldots Y_n$. Construct the squares $X_i X_{i+1} Y_{i+1} Y_i$, $i = 1, \ldots, n$ cyclically mod n. Join the diagonals of the squares X_i to Y_{i+1} sequentially to create the triangles. This defines B_n. It has $2n$ vertices, $2n$ triangular faces and two n-gon faces. The number of components is 3 if n is divisible by 3 and 1 otherwise. We call these doodles the **generalized Borromean** doodles. See Fig. D.2 for examples.

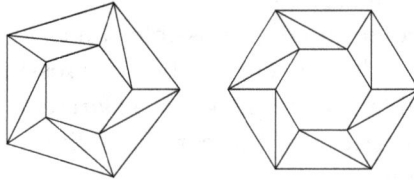

Fig. D.2. B_5 and B_6.

Exercise D.3. *Show that B_n is the closure of the twin $(t_1 t_2)^n$.*

Another two sequences can be defined by taking two concentric n-gons, $n > 2$, separated by a concentric $2n$-gon and filling in the annular regions with alternate squares and triangles. This can be done in two ways so that at the $2n$-gon each square or triangle in one annulus abuts a single square or triangle in the other annulus. So, taking nomenclature from the classification of polyhedra, we have **gyro**, C_n', which has the squares abutted by a common edge to the triangles while **ortho**, C_n'', has the squares abutted by a common edge to the squares and the triangles abutted to the triangles. Both C_n' and C_n'' have $4n$ vertices, $2n$ triangles and squares and two n-gons. The doodles C_n' and C_n'' for $n > 3$ can be distinguished by the number of components.

For $n = 3$, both C_3' and C_3'' have four components and a similar number of triangular and square regions. They are illustrated in Fig. D.3 and can be distinguished by the combinatorics of their regions.

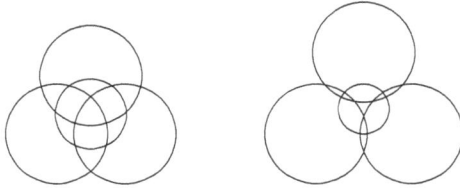

Fig. D.3. The gyro C_3' and the ortho C_3''.

In the table given later, these are $S12_1^4$ and $S12_2^4$.

We can describe C_3' and C_3'' as follows. Consider the Borromean doodle, B_3, with circular components. Now, draw a circle separating the innermost triangular region from the outermost triangular region. This describes C_3'. To obtain C_3'', perform an R_3 move on the innermost triangular region.

If we remove any component of C_3', we get the Borromean doodle. On the other hand, if we remove the innermost circle from C_3'', we get the trivial doodle. This is another proof that C_3' and C_3'' are distinct.

Lemma D.2.

(1) *The doodle C_n' has four components if n is divisible by 3. Otherwise, C_n' has two components.*
(2) *The doodle C_n'' has $n + 1$ components.*

Proof. Firstly, note that the central $2n$-gon is one of the components in both cases.

(1) Let the vertices around one of the n-gons be $P_1 \ldots P_n$. A component of C_n' containing the edge $P_i P_{i+1}$ contains the edge $P_{i+3} P_{i+4}$.
(2) For C_n'', each component which isn't the central $2n$-gon is a hexagon containing the edge $P_i P_{i+1}$, $i = 1, \ldots n \bmod n$, and there are n of these. $\qquad\square$

Exercise D.4. *Show that the doodle C'_n is the closure of the twin $(t_1 t_2 t_3 t_2)^n$. Can you find the twin whose closure is the doodle C''_n?*

D.4 Planar doodles and polyhedra

There is a bijection between minimal planar doodles and the 1-skeleta of 3-dimensional polyhedra whose vertices have valency four. It is well known that the Borromean doodle, B_3, is the 1-skeleton of the octahedron. In general, B_n is the 1-skeleton of the n-gon antiprism, see the work of Cromwell [28] for definitions.

Furthermore, C'_3 is the 1-skeleton of the cuboctahedron and C''_3 is the 1-skeleton of the anticuboctahedron or triangular orthobicupola which is Johnson's solid J_{27}. C'_4/C'_5 are also 1-skeleta of Johnson solids: J_{29}/J_{31}. C''_4/C''_5 are the 1-skeleta of the square gyrobicupola or pentagonal gyrobicupola and J_{28}/J_{30} are the 1-skeleta of the square orthobicupola/pentagonal orthobicupola. Further, bicupola for $n > 5$ cannot be realised with regular faces.

D.5 Amendments to doodles

There are a number of ways to alter a doodles diagram to produce a different doodle. In this section, we will consider three possibles.

D.5.1 *The ± 1 construction*

Let D be a doodle diagram on a surface Σ. Suppose that D has a region R with at least four edges and chose two disjoint edges e_1 and e_2 of R. Remove the interiors of the edges and join the dangling vertices with two diagonal arcs meeting at a new point X in the interior of R. This creates a new doodle diagram D'. We write $D \xrightarrow{+1} D'$ or $D' \xrightarrow{-1} D$ and call D an **ancestor** of D' and D' a **descendant** of D. The number of components of D' changes from that of D by 0 or ± 1, depending on how e_1 and e_2 are oriented and placed.

Lemma D.3. *Let D' have an ancestor D. Then D' is minimal if and only if D is minimal.*

Figure D.4 illustrates how the poppy is the ancestor of (the unique) minimal diagram with nine crossings and (the unique) two-component minimal diagram with 10 crossings.

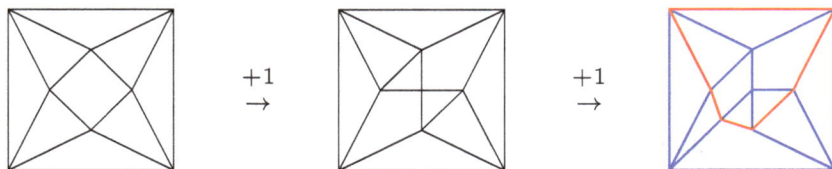

Fig. D.4. Descendants of the poppy.

A minimal diagram without ancestors is called an **orphan**. The following lemma gives a procedure for recognising whether a diagram is an orphan or not.

Lemma D.4. *If a minimal diagram has an ancestor, then there is a crossing X satisfying one of the following:*

(1) *There are two distinct regions appearing in a diagonal position about X such that they are a p'-gon and a q'-gon for some $p' > 3$ and $q' > 3$.*
(2) *There is a region appearing in a diagonal position about X such that it is an r'-gon for some $r' > 5$.*

Proof. Let D' be a descendant of D. Let e_1, e_2 and R be the edges and the region of D that were used to produce D'. (1) Consider a case that the regions containing e_1 and e_2, beside R, are distinct. Suppose they are p-gon and q-gon. Since D is minimal, $p > 2$ and $q > 2$. After applying the $+1$ construction, these regions become a p'-region and a q'-region, with $p' = p+1$ and $q' = q+1$. (2) Consider a case that the regions containing e_1 and e_2, beside R, are the same region of D. Suppose it is an r-gon. Since the boundary of this region contain e_1 and e_2, we have $r > 3$. By the $+1$ construction, this region becomes an r'-region with $r' = r + 2$. □

Theorem D.1. *The generalised Borromean doodles are all orphans.*

Proof. Their two regions with > 3 edges are protected by a ring of triangles. Thus, there is no crossing satisfying the condition of the previous lemma. □

D.5.2 *Enclosure*

Consider a non-minimal doodle diagram. Now, enclose one or more crossings with circles to make the new doodle minimal. The construction of the Borromean rings is a good example. Start with two circles meeting in two points. Now, enclose one of the crossings by a small circle. The doodle C_3'' illustrated in Fig. 9.3 is another example.

D.5.3 *Bridges*

If a doodle has a face with at least six edges, chose two edges which separate the face into p and q vertices with $p, q > 2$. Bridge the two edges creating two new faces with p and q edges. If the faces abutting the original two edges have i and j edges, then a new face with $i + j$ edges is created. The number of components either increases by one if the original edges belong to the same component or decreases the number of components by one otherwise. For examples of this see Fig. D.5.

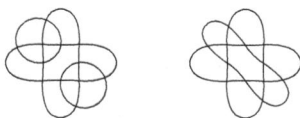

Fig. D.5. Enclosure and bridging.

Exercise D.5. *Invent spherical doodles, i.e. smooth immersions of spheres in 3-dimensional space in general position. Decide what conditions to place on the singularities and the allowable moves. Is there a unique minimal position?*

D.6 A table of prime and superprime doodles up to 14 crossings

A **prime doodle** is one where the removal of two crossings does not separate the remaining crossings. A **superprime doodle** is one where the removal of three crossings does not separate the remaining crossings.

The doodles have labels starting with P (prime) or S (super-prime), then the number of crossings, and a superscript that indicates the number of components. A subscript indicates the order which the computer in its wisdom has placed two doodles with the same previous indicators.

Planar doodles with less than 10 crossings:

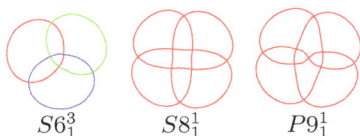

$$S6_1^3 \qquad S8_1^1 \qquad P9_1^1$$

Planar doodles with 10 crossings:

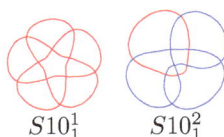

$$S10_1^1 \qquad S10_1^2$$

Planar doodles with 11 crossings:

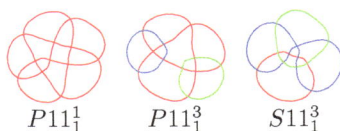

$$P11_1^1 \qquad P11_1^3 \qquad S11_1^3$$

Planar doodles with 12 crossings:

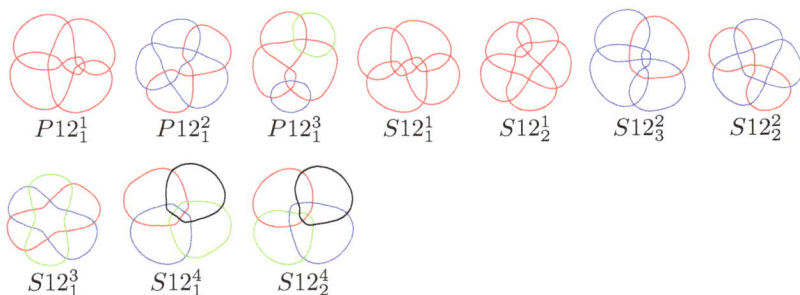

$$P12_1^1 \quad P12_1^2 \quad P12_1^3 \quad S12_1^1 \quad S12_2^1 \quad S12_3^2 \quad S12_2^2$$

$$S12_1^3 \qquad S12_1^4 \qquad S12_2^4$$

Planar doodles with 13 crossings:

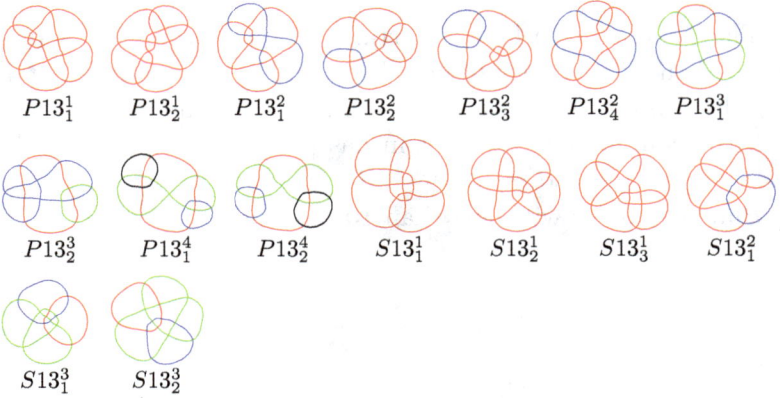

$P13_1^1$ $P13_2^1$ $P13_1^2$ $P13_2^2$ $P13_3^2$ $P13_4^2$ $P13_1^3$

$P13_2^3$ $P13_1^4$ $P13_2^4$ $S13_1^1$ $S13_2^1$ $S13_3^1$ $S13_1^2$

$S13_1^3$ $S13_2^3$

Planar doodles with 14 crossings:

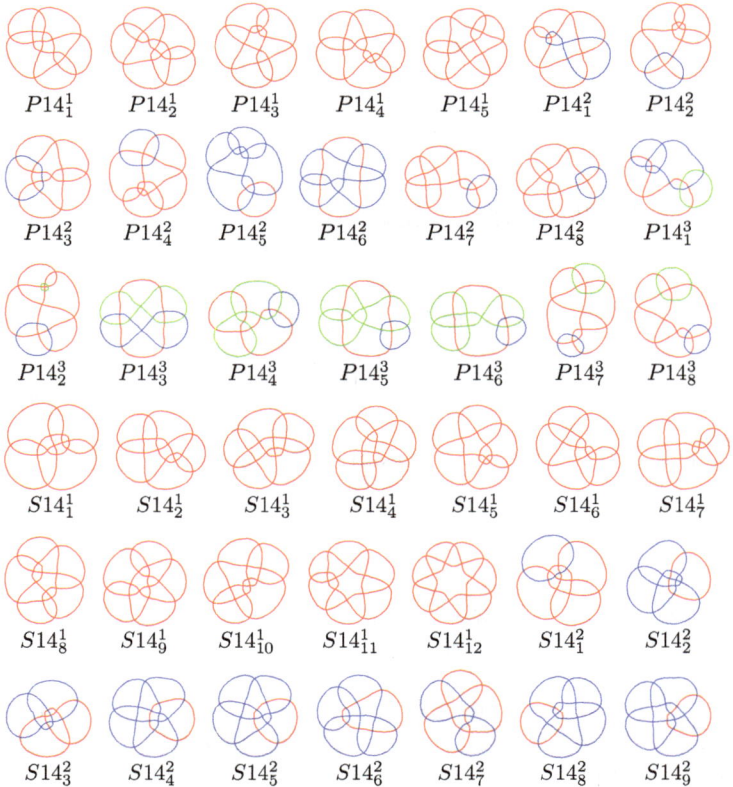

$P14_1^1$ $P14_2^1$ $P14_3^1$ $P14_4^1$ $P14_5^1$ $P14_1^2$ $P14_2^2$

$P14_3^2$ $P14_4^2$ $P14_5^2$ $P14_6^2$ $P14_7^2$ $P14_8^2$ $P14_1^3$

$P14_2^3$ $P14_3^3$ $P14_4^3$ $P14_5^3$ $P14_6^3$ $P14_7^3$ $P14_8^3$

$S14_1^1$ $S14_2^1$ $S14_3^1$ $S14_4^1$ $S14_5^1$ $S14_6^1$ $S14_7^1$

$S14_8^1$ $S14_9^1$ $S14_{10}^1$ $S14_{11}^1$ $S14_{12}^1$ $S14_1^2$ $S14_2^2$

$S14_3^2$ $S14_4^2$ $S14_5^2$ $S14_6^2$ $S14_7^2$ $S14_8^2$ $S14_9^2$

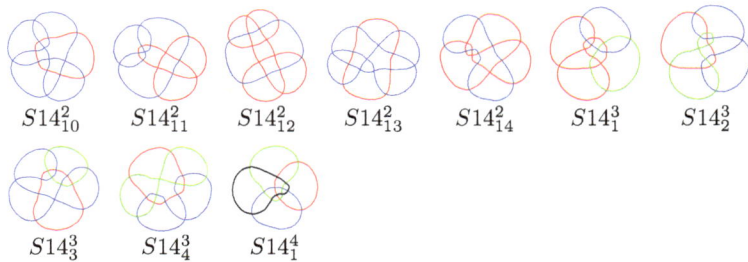

$S14_{10}^2$ $S14_{11}^2$ $S14_{12}^2$ $S14_{13}^2$ $S14_{14}^2$ $S14_1^3$ $S14_2^3$

$S14_3^3$ $S14_4^3$ $S14_1^4$

Bibliography

In this bibliography I have included all the references in the text and more. This collection is not complete and has no hope of being so. Every week I receive new details of papers published in this area. If anyone feels aggrieved that their result is not included please contact me and I will try to include it in the next edition if there is one.

1. J. W. Alexander, A lemma on a system of knotted curves. *Proc. Natl. Acad. Sci. USA.* 9, (1923), 93–95.
2. E. Artin, Theory of Braids. *E. Artin Annals of Mathematics, Second Series*, Vol. 48, No. 1 (1947), 101–126.
3. M. M. Asaeda, J. H. Przytycki and A. S. Sikora, Categorification of the Kauffman bracket skein module of I-bundles over surfaces. *Algebraic and Geometric Topology* **4**:52 (2004), 1177–1210.
4. R. S. Avdeev, On extreme coefficients of the Jones–Kauffman polynomial for virtual links. *J. Knot Theory Ramifications* **15**:7 (2006), 853–868.
5. V. G. Bardakov, The virtual and universal braids. *Fund. Math.* **184**, (2004), 1–18.
6. L. Bartholdi, B. Enriquez, P. Etingof and E. Rains, Groups and Lie algebras corresponding to the Yang–Baxter equations. *J. Algebra* **305**:2 (2006), 742–764.
7. andrewb@layer8.co.uk
8. A. Bartholomew and R. Fenn, Quaternionic invariants of virtual knots and links. *J. Knot Theory Ramifications* **17**:2 (2008), 231.
9. A. Bartholomew and R. Fenn, Biquandles of small size and some invariants of virtual and welded knots. *J. Knot Theory Ramifications* **20**:7 (2011), 943.
10. A. Bartholomew and R. Fenn, Planar doodles: Their properties, codes and classification. *J. To appear in J. Knot Theory Ramifications.*
11. A. Bartholomew and R. Fenn, Alexander and Markov theorems for generalized knots, I and II. *J. To appear in J. Knot Theory Ramifications.*

12. A. Bartholomew, R. Fenn, N. Kamada, S. Kamada and Doodles on surfaces *Journal of Knot Theory and Its Ramifications* **27**:12 (2018), 1850071.
13. A. Bartholomew, R. Fenn, N. Kamada and S. Kamada, New invariants of long virtual knots. *Kobe J. Math.* **27**:1–2 (2010), 21–33.
14. D. Bar-Natan, On Khovanov's categorification of the Jones polynomial. *Algebraic and Geometric Topology* **2**:16 (2000), 337–370.
15. D. Bar-Natan, Knot Atlas with Jeremey Green's atlas of virtual knots, http://www.math.toronto.edu/~drorbn/KAtlas/.
16. D. Bar-Natan, Khovanov's homology for tangles and cobordisms. *Geometry and Topology* **9–33** (2005), 1465–1499.
17. D. Bar-Natan and Z. Dancso, Finite type invariants of W-Knotted objects: From Alexander to Kashiwara and Vergne, http:/nlwww.math.toronto.edu /~drorbn/papers/WKO/,arXiv:math.GT/1309.7155.
18. George M. Bergman, The diamond lemma for ring theory. *Advances in Mathematics* **29** (1978), 178–218.
19. Quipu: Decorated permutation representations of finite groups. *Yongju Bae, J. Scott Carter, Byeorhi Kim*, to appear.
20. S. Bilson-Thompson, J. Hackett and L. H. Kauffman, Particle topology, braids, and braided belts. *J. Math. Phys.* **50**:11 (2009), 113505.
21. J. Bloom, Odd Khovanov homology is mutation invariant. *Math. Res. Lett.* **17**:1 (2010), 1–10.
22. A. Bouchet, Circle graph obstructions. *J. Combinatorial Theory Ser. B* **60** (1994), 107–144.
23. Tomas Boothby, Allison Henrich and Alexander Leaf, Minimal diagrams of free knots. J. Knot Theory Ramifications **23** (2014), 1450032.
24. M. O. Bourgoin, Twisted link theory. *Algebr. Geom. Topol.* **8**:3 (2008), 1249–1279.
25. S. Budden and R. Fenn, The equation $[B, (A-1)(A, B)] = 0$ and virtual knots and link. *Fund. Math.* **184** (2004), 19–29.
26. S. Budden and R. Fenn, Quaternion algebras and invariants of virtual knots and links II: The Hyperbolic Case. *J. Knot Theory Ramifications* **17** (2008), 305–314.
27. G. Burde and H. Zieschang, *Knots* de Gruyter, New York, 1985.
28. Peter Cromwell, *Polyhedra.* CUP, 1997.
29. J. S. Carter, D. Jelsovsky, S. Kamada and M. Saito, Quandle homology groups, their Betti numbers and virtual knots. *J. Pure Appl. Algebra* **157** (2001), 135–155.
30. J. S. Carter, S. Kamada and M. Saito, Geometric interpretations of quandle homology. *J. Knot Theory Ramifications* **10**:3 (2001), 345–386.
31. J. S. Carter, S. Kamada and M. Saito, Stable equivalence of knots on surfaces and virtual knot cobordisms. *J. Knot Theory Ramifications* **11**:3 (2002), 311–322.
32. J. S. Carter and M. Saito, Diagrammatic invariants of knotted curves and surfaces, Unpublished manuscript, 1992.
33. Zhiyun Cheng, A polynomial invariant of virtual knots, arXiv:math.GT/ 1202.3850.

34. M. W. Chrisman, A lattice of finite-type invariants of virtual knots. *Knots in Poland III, Part I*, **100** (2014), 27–49.

35. M. W. Chrisman, Knots in virtually fibered 3-manifolds, arXiv:math.GT/ 1405.6072v1.

36. M. W. Chrisman, On the Goussarov–Polyak–Viro finite-type invariants and the virtualization move. *J. Knot Theory Ramifications* **20**:3 (2011), 389–401.

37. M. W. Chrisman, Prime decomposition and non-commutativity in the monoid of long virtual knots, arXiv:math.GT/1311.5748.

38. M. W. Chrisman and H. A. Dye, The three loop isotopy and framed isotopy invariants of virtual knots. *Topology and Its Applications* **173**:15 (2014), 107–134.

39. M. W. Chrisman and V. O. Manturov, Fibered knots and virtual knots. *J. Knot Theory Ramifications*, **22**:12 (2013), 1341003, 23.

40. M. W. Chrisman and V. O. Manturov, Parity and exotic combinatorial formulae for finite-type invariants of virtual knots. *J. Knot Theory Ramifications* **21**:13 (2012), 1240001, 27.

41. K. Chu, Classification of flat virtual pure tangles. *J. Knot Theory Ramifications* **22**:4 (2013), 1340006.

42. J. Conant and P. Teichner, Grope cobordism of classical knots. *Topology* **43** (2004), 119–156.

43. P. Dehornoy, *Braids and Self-Distributivity*, Birkhauser, Basel, Vol. **192**, 2000.

44. C. H. Dowker and B. T. Morwen, Classification of knot projections. *Topology and Its Applications* **16**:1 (1983), 19–31.

45. H. Dye, *Characterizing Virtual Knots*, Ph.D. Thesis (2002).

46. H. A. Dye, Non-trivial realizations of virtual link diagrams, arXiv:math.GT/ 0502477.

47. H. A. Dye, Virtual knots undetected by 1 and 2-strand bracket polynomials, arXiv:math.GT/0402308.

48. H. A. Dye, A. Kaestner and L. H. Kauffman, Khovanov homology, Lee homology and a Rasmussen invariant for virtual knots (in preparation).

49. H. Dye and L. H. Kauffman, Minimal surface representations of virtual knots and links. *Algebr. Geom. Topol.* **5** (2005), 509–535.

50. H. Dye and L. H. Kauffman, Virtual knot diagrams and the Witten–Reshetikhin–Turaev invariant. *J. Knot Theory Ramifications* **14**:8 (2005), 1045–1075.

51. H. Dye and L. H. Kauffman, Virtual homotopy. *J. Knot Theory Ramifications* **19**:7 (2010), 935–960.

52. H. A. Dye and L. H. Kauffman, Virtual crossing number and the arrow polynomial. *J. Knot Theory Ramifications* **18**:10 (2009), 1335–1357.

53. H. A. Dye, L. H. Kauffman and V. O. Manturov, On two categorifications of the arrow polynomial for virtual knots. In *The Mathematics of Knots*, Contributions in the Mathematical and Computational Sciences, Vol. 1, Springer, Berlin, pp. 95–124, 2010.

54. S. Eliahou, L. Kauffman and M. Thistlethwaite, Infinite families of links with trivial Jones polynomial. *Topology* **42**, 155–169.

55. A. Fahmy, S. J. Glaser, L. H. Kauffman, S. J. Lomonaco, R. Marx, J. Myers, N. Pomplun and A. Spörl, NMR Quantum calculations of the Jones polynomial. *Phys. Rev.* **81** (2010), 03239.

56. R. Fenn, [www.maths.sussex.ac.ukStaff/RAF/Maths/historyi.jpg], ($i = 1, 2, \ldots$).

57. R. Fenn, *Techniques of Geometric Topology*. LMS Lect. Notes Series, Vol. 57, Cambridge University Press, Cambridge, 1983.

58. R. Fenn, Quaternion algebras and invariants of virtual knots and links I: The elliptic case. *J. Knot Theory Ramifications* **17** (2008), 279–304.

59. R. Fenn, Tackling the trefoils. *J. Knot Theory Ramifications* **21**:13 (2012), 1240004.

60. Roger Fenn, Biquandles for generalised knot theories. In *New Ideas in Low Dimensional Topology*, pp. 79–103 (2015) Generalised Series on Knots and Everything.

61. Roger Fenn, Gyo Taek Jin and Richárd Rimányi. Laces: A generalisation of braids. *Osaka J. Math.* **38** (2001), 251–269.

62. R. Fenn, M. Jordan and L. H. Kauffman, Biquandles and virtual links. *Topology and Its Applications* **145** (2004), 157–175.

63. R. A. Fenn, L. H. Kauffman and V. O. Manturov, Virtual knots: Unsolved problems. *Fundamenta mathematicae, Proceedings of the Conference "Knots in Poland-2003"*, **188** (2005), pp. 293–323.

64. R. Fenn, R. Rimanyi and C. Rourke, The braid permutation group. *Topology* **36**:1 (1997), 123–135.

65. R. Fenn and C. Rourke, Racks and links in codimension two. *J. Knot Theory Ramifications* **4** (1992), 343–406.

66. R. Fenn, C. Rourke and B. Sanderson, An introduction to species and the rack space. In *Topics in Knot Theory*, Springer, Dordrecht, 33–55, 1993.

67. R. Fenn, C. Rourke and B. Sanderson, Trunks and classifying spaces. *Applied Categorical Structures* **3** (1995), 321–356.

68. R. Fenn, C. Rourke and B. Sanderson, James bundles and applications, preprint, http:www.maths.warwick.ac.uk/cpr/ftp/james.ps (1996).

69. R. Fenn, C. Rourke and B. Sanderson, The rack space. *Trans. Amer. Math. Soc.* **359** (2007), 701–740.

70. R. Fenn and P. Taylor, Introducing doodles. *Lect. Notes in Maths. LMS*, Vol. 722, 37–43.

71. Roger Fenn, Ebru Keyman and Colin Rourke, The singular braid monoid embeds in a group, *J. Knot Theory Ramifications* **7**:7 (1998), 881–892.

72. R. Fenn and V. Turaev, Weyl algebras and knots. *J. Geometry and Physics* **57** (2007), 1313–1324.

73. A. T. Fomenko, The topology of surfaces of constant energy in integrable Hamiltonian systems, and obstructions to integrability. *Math. USSR Izv.* **29**:3 (1987), 629–658.

74. A. T. Fomenko, Topological invariants of Hamiltonian systems that are integrable in the sense of Liouville. *Funct. Anal. Appl.* **22**:4 (1988), 286–296.

75. A. T. Fomenko, The symplectic topology of completely integrable Hamiltonian systems. *Russian Math. Surveys* **44**:1 (1989), 181–219.

76. A. T. Fomenko, The theory of invariants of multidimensional integrable hamiltonian systems (with arbitrary many degrees of freedom). Molecular table of all integrable systems with two degrees of freedom. *Adv. Sov. Math.* **6** (1991), 1–35.

77. Konstantin Gotin, Markov theorem for doodles on two-sphere, arxiv. org/pdf/1807.05337.

78. M. Goussarov, M. Polyak and O. Viro, Finite type invariants of classical and virtual knots. *Topology* **39** (2000), 1045–1068.

79. M. Greene, *Some Results in Geometric Topology and Geometry*, Ph.D. Thesis, Warwick (1997).

80. M. Gutiérrez and S. Krstić, Normal forms for basis-conjugating automorphisms of a free group. *Int. J. Algebra Comput.* **8** (1998), 631–669.

81. N. Gügümcü and S. Lambropoulou, Knotoids, braidoids and applications. *Symmetry* **9** (2017), 315.

82. J. Hass and P. Scott, Shortening curves on surfaces. *Topology* **33**:1 (1994), 25–43.

83. D. Hrencecin, *On Filamentations and Virtual Knot Invariants*, Ph.D. Thesis (2001).

84. D. Hrencecin and L. H. Kauffman, On filamentations and virtual knots. *Topology and Its Applications* **134** (2003), 23–52.

85. Mikhail Khovanov, Doodle Groups. *Transactions of the American Mathematical Society* **349**:6 (June, 1997), 2297–2315.

86. D. P. Ilyutko, Framed 4-valent graphs: Euler tours, Gauss circuits and rotating circuits. *Sb. Math.* **202**:9 (2011), 1303–1326, *Mat. Sb.* **202**:9 (2011), 53–76 (in Russian).

87. D. P. Ilyutko, An equivalence between the set of graph-knots and the set of homotopy classes of looped graphs. *J. Knot Theory Ramifications* **21**:1 (2012), 1250001.

88. D. P. Ilyutko and V. O. Manturov, Introduction to graph-link theory. *J. Knot Theory Ramifications* **18**:6 (2009), 791–823.

89. D. P. Ilyutko and V. O. Manturov, Graph-links. *Dokl. Math.* **80**:2 (2009), 739–742, *Dokl. Akad. Nauk* **428**:5 (2009), 591–594 (in Russian).

90. D. P. Ilyutko and V. O. Manturov, Cobordisms of free knots. *Dokl. Math.* **80**:3 (2009), 844–846, *Dokl. Akad. Nauk* **429**:4 (2009), 439–441 (in Russian).

91. D. P. Ilyutko and V. O. Manturov, Graph-links. In *Introductory Lectures on Knot Theory, Selected Lectures Presented at the Advanced School and Conference on Knot Theory and Its Applications to Physics and Biology, Series of Knots and Everything*, Vol. 46, World Scientific, pp. 135–161, 2012.

92. D. P. Ilyutko, V. O. Manturov, and I. M. Nikonov, Virtual knot invariants arising from parities. *Knots in Poland III*, **100** (2013), 99–130.

93. D. P. Ilyutko, V. O. Manturov and I. M. Nikonov, Parity in knot theory and graph-links. *J. Math. Sci.* **193**:6 (2011), 809–965.

94. D. P. Ilyutko and V. S. Safina, Graph-links: Non-realizability, orientation and the Jones polynomial. *Topology, CMFD* **51** (2013), 33–63 (in Russian).

95. F. Jaeger, L. H. Kauffman and H. Saleur, The Conway polynomial in $\mathbb{R}^3$ and in thickened surfaces: A new determinant formulation. *J. Comb. Theory Ser. B* **61** (1994), 237–259.

96. V. F. R. Jones, A polynomial invariant for links via von Neumann algebras. *Bull. Amer. Math. Soc.* **129** (1985), 103–112.

97. V. F. R. Jones, Hecke algebra representations of braid groups and link polynomials. *Ann. of Math.* **126** (1987), 335–338.

98. V. F. R. Jones, On knot invariants related to some statistical mechanics models. *Pacific J. Math.* **137**:2 (1989), 311–334.

99. T. Kadokami, Detecting non-triviality of virtual links. *J. Knot Theory Ramifications* **12**:6 (2003), 781–803.

100. A. Kaestner and L. H. Kauffman, Parity, skein polynomials and categorification. *J. Knot Theory Ramifications* **21**:13 (2012), 1240011.

101. Seiichi Kamada, Braid presentation of virtual knots and welded knots, *Osaka J. Math.* **44**: 2 (2007), 441–458.

102. N. Kamada, Span of the Jones polynomial of an alternating virtual link. *Alg. and Geom. Topology* **4** (2004), 1083–1101.

103. N. Kamada, On the Jones polynomials of checkerboard colorable virtual knots, arXiv:math.PH/0008074.

104. N. Kamada and S. Kamada, Abstract link diagrams and virtual knots. *J. Knot Theory Ramifications* **9**:1 (2000), 93–106.

105. S. Kamada, Braid presentation of virtual knots and welded knots, arXiv:math.PH/0008092.

106. T. Kanenobu, Forbidden moves unknot a virtual knot. *J. Knot Theory Ramifications* **10**:1 (2001), 89–96.

107. L. H. Kauffman, *Formal Knot Theory*. Princeton University Press (1983), Dover Publications Inc. (2006).

108. L. H. Kauffman, State models and the Jones polynomial. *Topology* **26** (1987), 395–407.

109. L. H. Kauffman, *On Knots*, Princeton University Press, 1987.

110. L. H. Kauffman, Statistical mechanics and the Jones polynomial. *AMS Contemp. Math. Series* **78** (1989), 263–297.

111. L. H. Kauffman, Map coloring and the vector cross product. *J. Comb. Theo. B* **48**:2 (1990), 145–154.

112. L. H. Kauffman, *Knots and Physics,* World Scientific, Singapore–New Jersey –London–Hong Kong, 1991, 1994, 2001.

113. L. H. Kauffman, Map coloring, q-deformed spin networks, and Turaev–Viro invariants for 3-Manifolds. *Intl. J. Mod. Phys. B* **6**:11, 12 (1992), 1765–1794.

114. L. H. Kauffman, Gauss codes, quantum groups and ribbon Hopf algebras. *Reviews in Mathematical Physics* **5** (1993), 735–773. (Reprinted in 551–596).

115. L. H. Kauffman, Right integrals and invariants of three-manifolds. *Proceedings of Conference in Honor of Robion Kirby's 60th Birthday, Geometry and Topology Monographs*, Vol. 2 (1999), pp. 215–232.

116. L. H. Kauffman, Virtual knot theory. *European J. Comb.* **20** (1999), 663–690.

117. L. H. Kauffman, Introduction to virtual knot theory. *J. Knot Theory Ramifications* **21**:13 (2012), 1240007.

118. L. H. Kauffman, Knot theory and the heuristics of functional integration. *Physica A* **281** (2000), 173–200.

119. L. H. Kauffman, A survey of virtual knot theory. In *Proceedings of Knots in Hellas 98*, World Scientific (2000), pp. 143–202.

120. L. H. Kauffman, Detecting virtual knots. *Atti. Sem. Math. Fis., Univ. Modena*, Supplemento al vol. IL (2001), 241–282.

121. L. H. Kauffman, Vassiliev invariants and functional integration without integration. *Stochastic analysis and mathematical physics* (SAMP/ANESTOC 2002), pp. 91–114, World Scientific, River Edge, New Jersey, 2004.

122. L. H. Kauffman, A self-linking invariant of virtual knots. *Fund. Math.* **184** (2004), 135–158.

123. L. H. Kauffman, Non-commutative worlds. *New Journal of Physics* **6** (2004), 173.1–173.47.

124. L. H. Kauffman, Reformulating the map color theorem. *Discrete Math.* **302**:1–3 (2005), 145–172.

125. L. H. Kauffman, Knot diagrammatics. In *Handbook of Knot Theory edited by Menasco and Thistlethwaite*, Elsevier, Amsterdam, pp. 233–318, 2005.

126. L. H. Kauffman, An extended bracket polynomial for virtual knots and links. *J. Knot Theory Ramifications*, **18**:10 (2009), 1369–1422.

127. L. H. Kauffman, Non-commutative worlds and classical constraints, arXiv:math.PH/1109.1085.

128. L. H. Kauffman, Topological quantum information, Khovanov homology and the Jones polynonmial. In J. Colludo-Agustin and E. Hironaka, (ed.), *Topology of Algebraic Varieties and Singularities — Conference in Honor of Anatoly Libgober's 60-th Birthday*, Vol. 538 of *Contemporary Mathematics*, 245–264, AMS, American Mathematical Society, 2011.

129. L. H. Kauffman, Khovanov homology. In *Introductory Lectures on Knot Theory, Selected Lectures Presented at the Advanced School and Conference on Knot Theory and Its Applications to Physics and Biology, Series of Knots and Everything*, Vol. 46, World Scientific, Singapore, pp. 248–280, 2012.

130. L. H. Kauffman, An affine index polynomial invariant of virtual knots. *J. Knot Theory Ramifications* **22**:4 (2013), 1340007.

131. L. H. Kauffman, Diagrammatic knot theory (in preparation).

132. L. H. Kauffman, Virtual knot cobordism, arXiv:math.GT/1409.0324.

133. L. H. Kauffman, D. De Wit and J. Links, On the links-gould invariant of links. *J. Knot Theory Ramifications* **8**:2 (1999), 165–199.

134. L. H. Kauffman, Mo-Lin Ge and Y. Zhang, Yang-Baxterizations, universal quantum gates and Hamiltonians. *Quantum Inf. Process* **4**:3 (2005), 159–197.

135. L. H. Kauffman and S. Lambropoulou, Virtual braids. *Fund. Math.* **184** (2004), 159–186.

136. L. H. Kauffman and S. Lambropoulou, Virtual braids and the *L*-move. *J. Knot Theory Ramifications* **15**:6 (2006), 773–811.

137. L. H. Kauffman and S. Lambropoulou, Hard unknots and collapsing tangles. In *Introductory Lectures on Knot Theory, Selected Lectures Presented at the Advanced School and Conference on Knot Theory and Its Applications to Physics and Biology, Series of Knots and Everything*, Vol. 46, World Scientific, pp. 187–247, 2012.

138. L. H. Kauffman and S. Lambropoulou, A categorical model for the virtual braid group. *J. Knot Theory Ramifications* **21**:13 (2012), 1240008.

139. L. H. Kauffman and S. Lambropoulou, Low dimensional topology and modern physics (in preparation).

140. L. H. Kauffman and S. Lins, *Temperley–Lieb Recoupling Theory and Invariants of 3-Manifolds*. Princeton University Press, Princeton, 1994.

141. L. H. Kauffman and S. J. Lomonaco, Quantum entanglement and topological entanglement. *New J. Phys.* **4** (2002), 73.1–73.18.

142. L. H. Kauffman and S. J. Lomonaco, Quantum algorithms for the Jones polynomial. In Brandt, Donkor, Pirich, (eds.), *Quantum Information and Computation VIII – Spie Proceedings, April 2010, Vol. 7702, of Proceedings of Spie*, 770203-1 to 770203-11, SPIE 2010.

143. L. H. Kauffman and S. J. Lomonaco, Braiding operators are universal quantum gates. *New J. Phys.* **6** (2004), 134, 1–39.

144. L. H. Kauffman and S. J. Lomonaco, q-Deformed spin networks, knot polynomials and anyonic topological quantum computation. *J. Knot Theory Ramifications* **16**:3 (2007), 267–332.

145. L. H. Kauffman and S. J. Lomonaco, The Fibonaccii Model and the Temperley–Lieb Algebra. *International Journal Modern Phys. B* **22**:29 (2008), 5065–5080.

146. L. H. Kauffman and S. J. Lomonaco, Quantum knots. In Donkor Pirich and Brandt (eds.), *Quantum Information and Computation II – Proceedings of Spie, 12–14 April 2004* (2004), Intl. Soc. Opt. Eng, 268–284.

147. L. H. Kauffman and S. J. Lomonaco, Quantum knots and mosaics. *J. Quantum Information Processing* **7**:2-3 (2008), 85–115.

148. L. H. Kauffman and V. O. Manturov, Virtual knots and links. *Proc. Steklov Inst. Math.* **252** (2006), 104–121, *Tr. Mat. Inst. Steklova* **252** (2006), 114–133 (in Russian).

149. L. H. Kauffman and V. O. Manturov, Virtual biquandles. *Fund. Math.* **188** (2005), 103–146.

150. https://knotinfo.math.indiana.edu/.

151. L. H. Kauffman and D. E. Radford, Invariants of 3-manifolds derived from finite dimensional Hopf algebras. *J. Knot Theory Ramifications* **4**:1 (1995), 131–162.

152. L. H. Kauffman and D. E. Radford, Oriented quantum algebras and invariants of knots and links. *J. Algebra* **246** (2001), 253–291.

153. L. H. Kauffman and D. E. Radford, Oriented quantum algebras, categories and invariants of knots and links. *J. Knot Theory Ramifications* **10**:7 (2001), 1047–1084.

154. L. H. Kauffman and D. E. Radford, Bi-oriented quantum algebras, and a generalized alexander polynomial for virtual links. *Diagrammatic Morphisms and Applications* (San Francisco, CA, 2 000), 113–140, *Contemp. Math., 318, Amer. Math. Soc.,* Providence, RI, 2003.

155. L. H. Kauffman and H. Saleur, Free fermions and the Alexander–Conway polynomial. *Comm. Math. Phys.* **141** (1991), 293–327.

156. M. Khovanov, Doodle groups. *Trans. Amer. Math. Soc.,* **349**:6 (1997), 2297–2315.

157. M. Khovanov, A categorification of the Jones polynomial. *Duke Math. J* **101**:3 (1997), 359–426.

158. M. Khovanov, Link homology and Frobenius extensions. *Fund. Math* **190** (2006), 179–190.

159. M. Khovanov, A. D. Lauda, A diagrammatic approach to categorification of quantum groups I. *Represent. Theory* **13** (2009), 309–347.

160. M. Khovanov, L. Rozansky, Matrix factorizations and link homology. *Fund. Math.* **199**:1 (2008), 1–91.

161. M. Khovanov and L. Rozansky, Virtual crossings, convolutions and a categorification of the $SO(2N)$ Kauffman polynomial, arXiv:math.QA/0701333.

162. T. Kishino and S. Satoh, A note on non-classical virtual knots. *J. Knot Theory Ramifications* **13**:7 (2004), 845–856.

163. T. Kishino, 6 *kouten ika no kasou musubime no bunrui ni tsiuti (On classification of virtual links whose crossing number is less than or equal to 6),* Master Thesis (2000).

164. F. G. Korablev and S. V. Matveev, Reduction of knots in thickened surfaces and virtual knots. *Dokl. Akad. Nauk* **437**:6 (2011), 748–750.

165. V. A. Krasnov and V. O. Manturov, Graph-valued invariants of virtual and classical links and minimality problem. *J. Knot Theory Ramifications* **22**:12 (2013), 1341006, 14.

166. P. B. Kronheimer and T. S. Mrowka, Khovanov homology is an unknot-detector, arXiv:math.GT/1005.4346.

167. G. Kuperberg, What is a virtual link? *Algebr. Geom. Topol.* **3** (2003), 587–591.

168. S. Lambropoulou and C. Rourke Markov's theorem for 3-manifolds. *Topology and Its Applications,* **78** (1997) 95–112.

169. E. S. Lee, On Khovanov invariant for alternating links, arXiv:math.GT/0210213.

170. S. Y. Lee, Towards invariants of surfaces in 4-space via classical link invariants. *Trans. Amer. Math. Soc.* **361** (2009), 237–265.

171. W. B. R. Lickorish and K. C. Millett, Some evaluations of link polynomials. *Comment. Math. Helvetici* **61** (1986), 349–359.

172. W. B. R. Lickorish, A finite set of generators for the homotopy group of a 2-manifold, *Proc. Cambridge Philos. Soc.* **60** (1964), 769–778.

173. W. B. R. Lickorish, A representation of orientable combinatorial 3-manifolds. *Ann. Math.* **76**:2 (1962), 531–540.

174. R. Lipshitz, S. Sarkar and A Khovanov stable homotopy type. *J. Amer. Math. Soc.* **27**:4 (2014), 983–1042.

175. O. V. Manturov and V. O. Manturov, Free knots and groups. *J. Knot Theory Ramifications* **18**:2 (2009), 181–186.

176. O. V. Manturov and V. O. Manturov, Free Knots and Groups *Dokl. Akad. Nauk* **82**:2 (2010), 697–700, *Dokl. Akad. Nauk* **434**:1 (2010), 25–28 (in Russian).

177. V. O. Manturov, Atoms, height atoms, chord diagrams, and knots. Enumeration of atoms of low complexity using Mathematica 3.0. In *Topological Methods in Hamiltonian Systems Theory*, Moscow: Factorial (1998), pp. 203–212 (in Russian).

178. V. O. Manturov, The bracket semigroup of knots. *Math. Notes* **67**:4 (2000), 468–478, *Mat. Zametki* **67**:4 (2000), 549–562 (in Russian).

179. V. O. Manturov, Bifurcations, atoms and knots. *Moscow Univ. Math. Bull.* **55**:1 (2000), 1–7, *Vestnik Moskov. Univ. Ser. I Mat. Mekh* **1** (2000), 3–8 (in Russian).

180. V. O. Manturov, Invariants of virtual links. *Dokl. Math.* **65**:3 (2002), 329–331, *Dokl. Akad. Nauk* **384**:1 (2002), 11–13 (in Russian).

181. V. O. Manturov, Invariant two-variable polynomials for virtual links. *Russian Math. Surveys* **57**:5 (2002), 997–998, *Uspekhi Mat. Nauk* **57**:5(347) (2002), 141–142 (in Russian).

182. V. O. Manturov, On invariants of virtual links. *Acta Appl. Math.* **72**:3 (2002), 295–309.

183. V. O. Manturov, On recognition of virtual braids. *Math. Sci. J. (New York)* **131**:1 (2003), 5409–5419, *Geom. Topol.* **8**, *Zap. Nauchn. Sem. POMI* **299** (2003), 267–286 (in Russian).

184. V. O. Manturov, Curves on surfaces, virtual knots, and the Jones–Kauffman polynomial. *Dokl. Math.* **67**:3 (2003), 326–328, *Dokl. Akad. Nauk* **390**:2 (2003), 155–157 (in Russian).

185. V. O. Manturov, Atoms and minimal diagrams of virtual links. *Dokl. Math.* **68**:1 (2003), 37–39, *Dokl. Akad. Nauk* **391**:2 (2003), 166–168 (in Russian).

186. V. O. Manturov, Combinatorial problems in virtual knot theory. *Math. Problems Cybernetics* **12** (2003), 147–178 (in Russian).

187. V. O. Manturov, Multivariable polynomial invariants for virtual knots and links. *J. Knot Theory Ramifications* **12**:8 (2003), 1131–1144.

188. V. O. Manturov, Kauffman-like polynomial and curves in 2-surfaces. *J. Knot Theory Ramifications* **12**:8 (2003), 1145–1153.

189. V. O. Manturov, *Knot Theory*, CRC-Press, Boca Raton, 2004.

190. V. O. Manturov, Invariant polynomials of virtual links. *Tr. Mosk. Mat. Obs.* **65**:1 (2004), 175–200 (in Russian).

191. V. O. Manturov, Finite-type invariants of virtual links and the Jones–Kauffman polynomial. *Dokl. Math.* **69**:2 (2004), 164–166, *Dokl. Akad. Nauk* **395**:1 (2004), 18–21 (in Russian).

192. V. O. Manturov, The Khovanov polynomial for virtual knots. *Dokl. Math.* **70**:2 (2004), 679–681, *Dokl. Akad. Nauk* **398**:1 (2004), 15–18 (in Russian).

193. V. O. Manturov, Long virtual knots and their invariants. *J. Knot Theory Ramifications* **13**:8 (2004), 1029–1039.

194. V. O. Manturov, Virtual knots and infinite-dimensional Lie algebras. *Acta Appl. Math.* **83** (2004), 221–233.

195. V. O. Manturov, *Teoriya uzlov* (*Knot theory*). RCD, Moscow–Izhevsk, 2005 (in Russian).

196. V. O. Manturov, On long virtual knots. *Dokl. Math.* **71**:2 (2005), 253–255, *Dokl. Akad. Nauk* **401**:5 (2005), 595–598 (in Russian).

197. V. O. Manturov, The Khovanov complex for virtual links. *J. Math. Sci.* (*New York*) **144**:5 (2005), 4451–4467, *Fundam. Prikl. Mat.* **11**:4 (2005), 127–152 (in Russian).

198. V. O. Manturov, A proof of Vassiliev's conjecture on the planarity of singular links. *Izv. Math.* **69**:5 (2005), 1025–1033, *Izvestiya RAN, Ser. Mat.* **69**:5 (2005), 169–178 (in Russian).

199. V. O. Manturov, Vassiliev invariants for virtual links, curves on surfaces and the Jones–Kauffman polynomial. *J. Knot Theory Ramifications* **14**:2 (2005), 231–242.

200. V. O. Manturov, Flat hierarchy. *Fund. Math.* **188** (2005), 147–154.

201. V. O. Manturov, The Khovanov complex and minimal knot diagrams. *Dokl. Math.* **73**:1 (2006), 46–48, *Dokl. Akad. Nauk* **406**:3 (2006), 308–311 (in Russian).

202. V. O. Manturov, Khovanov homology of virtual knots with arbitrary coefficients. *Izv. Math.* **71**:5 (2007), 967–999, *Izvestiya RAN, Ser. Mat.* **71**:5 (2007), 111–148 (in Russian).

203. V. O. Manturov, Khovanov homology for virtual links with arbitrary coefficients. *J. Knot Theory Ramifications* **16**:3 (2007), 345–377.

204. V. O. Manturov, Additional gradings in Khovanov's complex for thickened surfaces. *Dokl. Math.* **77**:3 (2008), 368–370, *Dokl. Akad. Nauk* **420**:2 (2008), 168–171 (in Russian).

205. V. O. Manturov, Additional gradings in Khovanov homology. In *Topology and Physics. Dedicated to the Memory of X.-S. Lin*, Nankai Tracts in Mathematics, World Scientific, Singapore, pp. 266–300, 2008.

206. V. O. Manturov, Compact and long virtual knots. *Tr. Mosk. Mat. Obs.* **69** (2008). 5–33.

207. V. O. Manturov, Embeddings of 4-valent framed graphs into 2-surfaces. *Dokl. Math.* **79**:1 (2009), 56–58, *Dokl. Akad. Nauk* **424**:3 (2009), 308–310 (in Russian).

208. V. O. Manturov, On free knots, arXiv:math.GT/0901.2214.

209. V. O. Manturov, On free knots and links, arXiv:math.GT/0902.0127.

210. V. O. Manturov, Free knots are not invertible, arXiv:math.GT/0909.2230v2.

211. V. O. Manturov, Parity and cobordisms of free knots, arXiv:math.GT/1001.2728.

212. V. O. Manturov, Parity in knot theory. *Sb. Math.* **201**:5 (2010), 693–733, *Mat. Sb.* **201**:5 (2010), 65–110 (in Russian).

213. V. O. Manturov, Embeddings of four-valent framed graphs into 2-surfaces. In *The Mathematics of Knots*, Contributions in the Mathematical and Computational Science, Vol. 1 (2010), pp. 169–197.

214. V. O. Manturov, Parity and cobordisms of free knots. *Mat. Sb.* **203**:2 (2012), 45–76 (in Russian).

215. V. O. Manturov, Free knots and parity. In *Introductory Lectures on Knot Theory, Selected Lectures Presented at the Advanced School and Conference on Knot Theory and Its Applications to Physics and Biology, Series of Knots and Everything*, Vol. 46, World Scientific, pp. 321–345, 2012.

216. V. O. Manturov, Free knots, groups and finite–order invariants, Statu Nascendi. https://arxiv.org/abs/1004.4325.

217. V. O. Manturov and D. P. Ilyutko, *Virtual Knots: The State of the Art.* World Scientific, Singapore, 2013.

218. A. A. Markov, ber die freie Äquivalenz geschlossener Zöpfe. *Recueil Mathématique Moscou* **1** (1935).

219. W. S. Massey, *Algebraic Topology: An Introduction*, Springer, New York.

220. S. Matveev and C. Hog-Angeloni, Roots in 3-manifold topology. *Geometry and Topology Monographs* **14** (2008), 295–319.

221. W. Menasco and M. B. Thistlethwaite, The classification of alternating links. *Ann. Math.* **138** (1993), 113–171.

222. Moise E. E. (1977) Geometric Topology in Dimensions 2 and 3. In *Graduate Texts in Mathematics*, Vol. 47. Springer, New York, NY.

223. S. Morrison and A. Nieh, On Khovanov's cobordism theory for $su(3)$ knot homology, arXiv:math.GT/0612754.

224. H. R. R. Morton, Threading knot diagrams., *Math. Proc. Camb. Phil. Soc.* 99 (1986), 247–260.

225. K. Murasugi, The Jones polynomial and classical conjectures in knot theory. *Topology* **26** (1987), 187–194.

226. S. Nelson, Unknotting virtual knots with Gauss diagram forbidden moves. *J. Knot Theory Ramifications* **10**:6 (2001), 931–935.

227. S. Nelson, Virtual crossing realization, arXiv:math.GT/0303077.

228. M. H. A. Newman. On theories with a combinatorial definition of "equivalence." *Ann. Math.*, **43**: 2, (1942), 223–243.

229. M. Niebrzydowski, J. H. Przytycki, Homology of dihedral quandles. *J. Pure Appl. Algebra* **213**:5 (2009), 742–755.

230. I. Nikonov, Khovanov homology of graph-links, arXiv:math.GT/1005.2812.

231. I. Nikonov, Odd Khovanov homology of principally unimodular bipartite graph-links, arXiv:math.GT/1006.0161.

232. P. Ozsváth, J. Rasmussen and Z. Szabó, Odd Khovanov homology, arXiv:math.QA/0710.4300.

233. P. Ozsváth and Z. Szabó, Heegard Floer homology and alternating knots. *Geometry and Topology* **7** (2003), 225–254.

234. J. A. Rasmussen, Khovanov homology and the slice genus, arXiv:math.GT/0402131.

235. K. Reidemeister, Elementare Begründung der Knotentheorie. *Abh. Math. Sem. Univ. Hamburg*, 5(1): (1927), 24–32.

236. Rolfsen, Dale, *Knots and Links*. AMS Chelsea Publishing, Rhode Island, 2004.
237. C. Rourke, What is a welded link? In *Intelligence of low dimensional topology, Series of Knots and Everything*, Vol. 40, World Scientific, pp. 263–270, 2006.
238. S. Sarkar, Grid diagrams and shellability, arXiv:math.GT/0901.2156v3.
239. S. Satoh, Virtual knot presentation of ribbon torus-knots. *J. Knot Theory Ramifications* **9**:4 (2000), 531–542.
240. J. Sawollek, On Alexander-Conway polynomials for virtual knots and links, arXiv:math.GT/9912173.
241. J. Sawollek, An orientation-sensitive Vassiliev invariant for virtual knots, arXiv:math.GT/0203123v3.
242. W. J. Schellhorn, Filamentations for virtual links, arXiv:math.GT/0402162.
243. H. Seifert, Über das Geschlecht von Knoten. *Mathematische Annalen* (1935) 571–592.
244. S. Stephen, A classification of immersions of the two-sphere. *Transac. Am. Mathe. Soc.* **90**:2 (1959). 281–290.
245. D. S. Silver and S. G. Williams, Alexander groups and virtual links. *J. Knot Theory Ramifications* **10** (2001), 151–160.
246. D. S. Silver and S. G. Williams, Alexander groups of long virtual knots, arXiv:math.GT/0405460.
247. D. S. Silver and S. G. Williams, On a class of virtual knots with unit Jones polynomial. *J. Knot Theory Ramifications* **13**:3 (2004), 367–371.
248. D. S. Silver and S. G. Williams, Polynomial invariants of virtual links. *J. Knot Theory Ramifications* **12**:7 (2003), 987–1000.
249. D. Stanovský, On axioms of biquandles. *J. Knot Theory Ramifications* **15**:7 (2006), 931–933.
250. F. Swenton, On a calculus for surfaces and 2-knots in 4-space. *J. Knot Theory Ramifications* **10**:8 (2001), 1133–1141.
251. Y. Takeda, Introduction to virtual surface-knot theory. *J. Knot Theory Ramifications* **21**:14 (2012), 1250131.
252. M. Thistlethwaite, Links with trivial Jones polynomial. *J. Knot Theory Ramifications* **10**:4 (2001), 641–643.
253. M. B. Thistlethwaite, A spanning tree expansion of the Jones polynomial. *Topology* **26**:3 (1987), 297–309.
254. D. B. A. Epstein, James W. Cannon *et al.* (eds.) *Word Processing in Groups*. Jones and Bartlett Publishers, Boston, MA, 1992.
255. L. Traldi and L. Zulli, A bracket polynomial for graphs. *J. Knot Theory Ramifications* **18** (2009), 1681–1709.
256. Paweł Traczyk, *A New Proof of Markov's Braid Theorem*. Banach Center Publications, Vol. 42, Iss. 1, pp. 409–419, 1998.
257. V. Turaev, Virtual strings and their cobordisms, arXiv:math.GT/0311185.
258. V. Turaev, Cobordism of knots on surfaces. *J. Topol.* **1**:2 (2008), 285–305.
259. V. Turaev, Cobordisms of words. *Commun. Contemp. Math.* **10**:1 (2008), 927–972.
260. V. Turaev, Knotoids. *Osaka J. Math.* 49 (2012), 195–223.

261. V. G. Turaev and P. Turner, Unoriented topological quantum field theory and link homology. *Algebr. Geom. Topol.* **6** (2006), 1069–1093.

262. V. V. Vershinin, On homology of virtual braids and Burau representation, arXiv:math.GT/9904089.

263. O. Viro, Khovanov homology, its definitions and ramifications. *Fund. Math.* **184** (2004), 317–342.

264. P. Vogel, Representation of links by braids: A new algorithm. *Comment. Math. Helvetici* **65** (1990), 104–113.

265. B. Webster, Knot invariants and higher representation theory I: diagrammatic and geometric categorification of tensor products, arXiv:math.GT/1001.2020.

266. B. Webster, Knot invariants and higher representation theory II: the categorification of quantum knot invariants, arXiv:math.GT/1005.4559.

267. Whitney, H. On regular closed curves on the plane. *Compositio Math.* **4** (1937), 276–284.

268. S. Winker, *Quandles, Knots Invariants and the N-Fold Branched Cover*, Ph.D. Thesis (1984).

269. E. Witten, Quantum field theory and the Jones polynomial. *Comm. Math. Phys.* **121** (1989), 351–399.

270. S. Yamada, The minimal number of Seifert circles equals the braid index of a link, *Invent. Math.* **89** (1987), 347–356.

271. P. Zinn-Justin and J. B. Zuber, Matrix integrals and the generation and counting of virtual tangles and links. *J. Knot Theory Ramifications* **13**:3 (2004), 325–355.

272. P. Zinn-Justin and J. B. Zuber, Tables of Alternating Virtual Knots, http://ipnweb.in2p3.fr/~lptms/membres/pzinn/virtlinks/.

Index